Praise for *All That Remains*

Book of the Year, Saltire Literary Awards
A CrimeReads Best True Crime Book of the Month

"With a disarming frankness . . . a multipronged approach to the topic of death, exploring it through scientific, sociological, historical, and philosophical lenses. . . . This is a perceptive study of a subject both deeply uncomfortable and uncommonly engrossing."

—*Publishers Weekly*

"Black's testimony to the nobility of her calling is a welcome and compassionate look at death and the mysteries that shroud it."

—*Booklist*

"Essential . . . an insightful, compelling, and often entertaining memoir about a life spent studying and reckoning with the dead and their secrets."

—*CrimeReads*, "The Best True-Crime Books of the Month"

"Dame Sue Black . . . writes vividly about her job identifying human remains, the events in her life that led her to this career, and the reality of death in all of our lives."

—*Book Riot*, "33 Highly Anticipated Crime Novels"

"Dame Sue Black writes about life and death with great tenderness but no nonsense, with impeccable science lucidly explained, and with moral depths humanely navigated, so that we can all feel better about the path we must all inevitably follow."

—Lee Child

"*All That Remains* provides a fascinating look at death—its causes, our attitudes toward it, the forensic scientist's way of analyzing it. A unique and thoroughly engaging book."

—Kathy Reichs, *New York Times* bestselling author

"No scientist communicates better than Professor Sue Black. *All That Remains* is a unique blend of memoir and monograph that admits us into the remarkable world of forensic anthropology."

—Val McDermid, award-winning and bestselling author

"Most of us are terrified of death, but Sue Black shows us that death is in fact a wondrous process, intimately tied with life itself. Written with warmth and humanity, *All That Remains* reveals her life among the dead, who can surely count her as their best friend."

—Tess Gerritsen, internationally bestselling author

"This fascinating look by a world-leading forensic scientist at what the dead can tell us is a real eye-opener. . . . Part meditation, part popular science, and part memoir . . . the book offers a close-up and startlingly clear view of a subject that makes most of us look away. . . . Extraordinary."

—*Sunday Times*

"An engrossing memoir . . . an affecting mix of the personal and professional."

—*Financial Times*

"A model of how to write about the effect of human evil without losing either objectivity or sensitivity. . . . Heartening and anything but morbid . . . Leaves you thinking about what kind of human qualities you value, what kinds of people you actually want to be with."

—*New Statesman*

"Black's utterly gripping account of her life and career as a professor of anatomy and forensic anthropology manages to be surprisingly life-affirming. As she herself says, it is 'as much about life as about death.'"

—*Guardian*

"The real thing here is not the cause of death, but the nature of the life. Black is genuinely moving about the respect we should have for the dead. . . . There is much to admire in this book."

—*Scotsman*

"For someone whose job is identifying corpses, Sue Black is a cheerful soul. . . . *All That Remains* feels like every episode of *Silent Witness*, pre-fictionalized. Except, you know, really good."

—*The Times*

"Poignant and thought-provoking . . . It is the book's humanity which will connect with readers."

—*Scottish Daily Mail*

"Let [Sue Black] take you by the hand and lead you on a journey which will inspire your awe and devotion. . . . A wonderful surprise of a book."

—*The Tablet*

"Sue Black has been intimately involved with the aftermath of death for her whole professional career, and in her book she weaves in details of her amazing and active life with her analysis of death in a narrative that is personal, touching, occasionally tragic, but also instilled with her wonderful sense of humour."

—Dr. Richard Shepherd, consultant forensic pathologist

"This is one of those books that'll astound as it entertains. It's a little shivery and oh-so-fascinating. And in the end, *All that Remains* is a tale you can live with."

—*Marco Eagle*, part of the *USA Today* network

"Black is informative, respectful, easily accessible, and funny. This is perfect for anatomy nerds and CSI fans."

—*bethfishreads.com*

ALL *that* REMAINS

A RENOWNED FORENSIC SCIENTIST ON DEATH, MORTALITY, AND SOLVING CRIMES

SUE BLACK

ARCADE PUBLISHING • NEW YORK

For Tom, for ever my love and my life.
And for Beth, Grace and Anna—each is my favourite daughter.
Thank you all for making every moment of my life worthwhile.

First North American Paperback Edition 2020

First published in Great Britain by Doubleday, an imprint of Transworld Publishers.

Arcade Publishing books may be purchased in bulk at special discounts for sales promotion, corporate gifts, fund-raising, or educational purposes. Special editions can also be created to specifications. For details, contact the Special Sales Department, Arcade Publishing, 307 West 36th Street, 11th Floor, New York, NY 10018 or arcade@skyhorsepublishing.com.

Arcade Publishing® is a registered trademark of Skyhorse Publishing, Inc.®, a Delaware corporation.

Visit our website at www.arcadepub.com.

10

Library of Congress Cataloging-in-Publication Data is available on file.
Library of Congress Control Number: 2018965129

Cover design by Erin Seaward-Hiatt

Paperback ISBN: 978-1-950691-91-3
Hardcover ISBN: 978-1-948924-27-6
Ebook ISBN: 978-1-94892-429-0

Printed in the United States of America

CONTENTS

Introduction

'Death is not the greatest loss in life.
The greatest loss is what dies inside us while we live'

Norman Cousins, political journalist (1915–1990)

Me at about two years of age.

Death and the hyped-up circus that surrounds her are perhaps more laden with clichés than almost any other aspect of human existence. She is personified as sinister, as a harbinger of pain and unhappiness; a predator who haunts and hunts from the shadows, a dangerous thief in the night. We give her ominous and cruel nicknames—the Grim Reaper, the Great Leveller, the Dark Angel, the Pale Rider—and portray her as a gaunt skeleton in a dark, hooded cloak wielding a deadly scythe, destined to separate our soul from our body with one lethal swipe. Sometimes she is a black, feathered spectre that hovers menacingly over us, her cowering victims. And, despite being feminine in many languages where nouns have genders (including Latin, French, Spanish, Italian, Polish, Lithuanian and Norse), she is often none the less depicted as a man.

It is easier to treat death unkindly because in the modern world she has become a hostile stranger. For all the progress humanity has made, we are little closer to deciphering the complex bonds between life and death than we were hundreds of years ago. Indeed, in some respects, we are perhaps further away than ever before from understanding her. We seem to have forgotten who death is, what her purpose is, and, where our ancestors perhaps considered her a friend, we choose to treat her as an unwelcome and devilish adversary to be avoided or bested for as long as possible.

Our default position is either to vilify or to deify death, sometimes vacillating between the two. Either way, we prefer not to mention her if we can help it in case it encourages her to come too close. Life is light, good and happy; death is dark, bad and sad. Good and evil, reward and punishment, heaven and hell, black and white—our Linnaean tendencies lead us to neatly categorise life and death as opposites, giving us the comforting illusion of an unambiguous sense of right and wrong that perhaps unfairly banishes death to the dark side.

As a result we have come to dread her presence as if she were somehow infectious, afraid that if we attract her attention then she might come for us before we are remotely ready to stop living. We may conceal our fear by putting on a show of bravado or poking fun at her in the hope of anaesthetising ourselves to her sting. We know, though, that we will not be laughing when we reach the top of her list and she does finally call out our name. So we learn at a very early age to be hypocritical about her, ridiculing her with one turn of the face and then becoming deeply reverential with another. We learn a new language to try to blunt her sharp edges and dull the pain. We talk about 'losing' someone, whisper of their 'passing' and, in sombre respectful tones, we commiserate with others when a loved one has 'gone'.

I didn't 'lose' my father—I know exactly where he is. He is buried at the top of Tomnahurich Cemetery in Inverness, in a lovely wooden box provided by Bill Fraser, the family funeral director, of which he might have approved, although he would probably have thought it too expensive. We put him in a hole in the ground on top of the disintegrating coffins of his mother and father, neither of which will now hold more than their bones and the few teeth they still had when they died. He has not passed, he is not gone, he is not lost: he is dead. Indeed, he better not have gone anywhere—that would be most troublesome and inconsiderate of him. His life is extinct, and none of the euphemistic rhetoric in the world will ever bring it, or him, back.

As the product of a strict, no-nonsense, Scottish Presbyterian family where a spade was called a shovel and empathy and sentimentality were often viewed as weaknesses, I like to think my upbringing has made me pragmatic and thick-skinned, a coper and a realist. When it comes to matters of life and death I harbour no misconceptions and in discussing them I try to be honest and truthful, but that does not mean I don't care, and it doesn't make me immune to pain and grief or unsympathetic to that of others. What I do not have is a maudlin sentimentality about death and the dead. As Fiona, our inspirational chaplain at Dundee University, puts it so eloquently, there is no comfort to be had from soft words spoken at a safe distance.

With all our twenty-first-century sophistication, why do we still opt to take cover behind familiar, safe walls of conformity and denial, rather than opening up to the idea that maybe death is not the demon we fear? She does not need to be lurid, brutal or rude. She can be silent, peaceful and merciful. Perhaps the answer is that we don't trust her because we don't choose to get to know her, to take the trouble in the course of our lives to try to understand her. If we did, we might learn to accept her as an integral and fundamentally necessary part of our life's process.

We view birth as the beginning of life and death as its natural end. But what if death is just the beginning of a different phase of existence? This, of course, is the premise of most religions, which teach that we should not fear death as it is merely the gateway to a better life beyond. Such beliefs have brought solace to many through the ages, and perhaps the vacuum left by the increasing secularisation of our society has contributed to the resurgence of an ancient, instinctive but unsubstantiated aversion to death and all its trappings.

Whatever we believe, life and death are unquestionably inextricably bound parts of the same continuum. One does not, and cannot, exist without the other and, no matter how much modern medicine strives to intervene, death will ultimately prevail. Since there is no way we can ultimately prevent it, perhaps our time would be better spent focusing on improving and savouring the period between our birth and our death: our life.

Herein lies one of the fundamental differences between forensic pathology and forensic anthropology. Forensic pathology seeks evidence of a cause and manner of death—the end of the journey—whereas forensic anthropology reconstructs the life led, the journey itself, across the full span of its duration. Our job is to reunite the identity constructed during life with what remains of the corporeal form in death. So forensic pathology and anthropology are partners in death and, of course, in crime.

In the UK, anthropologists, unlike pathologists, are scientists rather than doctors and are therefore unlikely to be medically qualified to certify a death or the cause of death. In these days of ever-expanding scientific knowledge, pathologists cannot be expected to be experts in everything, and the anthropologist has an important role to play in the investigation of serious crimes involving a death. Forensic anthropologists assist in unravelling the clues associated with the identity of the victim and may aid the pathologist to reach his or her final decisions about the manner and cause of death. Each discipline brings its own complementary and specific skills to the mortuary table.

On one such mortuary table, for example, a pathologist and I were faced with human remains in an advanced stage of decomposition. The skull was shattered into over forty jumbled fragments. As the medically qualified practitioner, her remit was to determine the cause of death and she was pretty sure it was going to be gunshot injury. But she needed to be certain. Surveying with dismay the multitude of fragments of white bone on the grey metal tabletop, she said, 'I can't identify all the pieces, let alone try to stick them together. That's your job.'

The forensic anthropologist's role is first to help establish who the person may have been in life. Were they male or female? Tall or short? Old or young? Black or white? Does the skeleton show evidence of any injuries or disease that might be linked to medical or dental records? Can we extract information from bones, hair and nails about their composition which might tell us where the person was living and the type of food they ate? And in this case, could we undertake a three-dimensional human jigsaw puzzle to allow us to reveal not only the cause of death, which was indeed gunshot injury to the skull, but also the manner of death? By gathering this information and completing the jigsaw we were able to establish the identity of the young man and to corroborate eyewitness testimony by confirming a ballistic entry wound to the back of the head and its exit from the forehead above and between his eyes. This was a close-range execution, in which the victim had been kneeling when the firearm was placed directly against the skin at the back of his head. He was just fifteen and his crime was his religion.

Another illustration of the symbiotic relationship between anthropologist and pathologist concerned an unfortunate young man who was beaten to death after confronting a group of youths intent on vandalising a car in the street outside his house. His body had been kicked and punched, he had suffered fatal impact trauma to his head and he exhibited multiple skull fractures. In this case we knew the identity of the victim, and the pathologist was able to determine the cause of death as blunt-force trauma, resulting in massive internal haemorrhage. But she also wanted to report on how his death was brought about and, in particular, on the type of implement most likely to have been used to kill him. We were able to identify every fragment of the skull and to reconstruct it, enabling the pathologist to confirm that there had been one primary blow to the head, made by a hammer, or something of similar shape, which had caused a focal depressed fracture and multiple radiating fractures leading to the fatal intracranial bleeding.

For some, the distance between the beginning and the end of life will be lengthy, perhaps over a century, whereas for others, like these murder victims, the two events will occur much closer together. Sometimes they may be separated only by a fleeting but precious few seconds. From the point of view of the forensic anthropologist, a long life is good news, as the longer it has been, the more scars of experience will be written and stored within the body, and the clearer their imprint on our mortal remains will be. For us, unlocking this information is almost like reading it in a book, or downloading it from a USB stick.

In the eyes of most people, the worst outcome of this earthly adventure is a life cut short. But who are we to judge what is short? What is not in doubt is that the longer we survive beyond birth, the higher the probability will be that our lives will end sooner rather than later: we are more likely, in most cases, to be closer to death at ninety than we are at twenty. And logic tells us that we will never again be further away from a personal acquaintance with death than we are right at this moment.

So why are we surprised when people die? Over 55 million of us around the world do it every year—two a second—and it is the one event of our lives that we know with absolute certainty is going to happen to every single one of us. This by no means diminishes our sadness and grief when it happens to someone close to us, of course, but its inevitability demands an approach that is both practical and realistic. Since we can't influence the creation of our lives, and their end is unavoidable, perhaps we should be focusing on what we can regulate: our expectations of the distance between them. Perhaps it is this we should be trying to manage more effectively by measuring, acknowledging and celebrating its value rather than its duration.

In the past, when death was less easy to postpone, we may have been better at this. In Victorian times, for example, when infant mortality was high, nobody was surprised if a child did not reach its first birthday. Indeed, it was not unusual for several children in a family to be given the same name to ensure that it survived, even if the child did not. In the twenty-first century, infant death is more shocking, but to be stunned when someone dies at the age of ninety-nine defies all logic.

Society's expectations are the battleground of every medical expert who aims to force death into a retreat. The best they can hope to do is to buy more time and expand the distance between our two mortal events. That they will ultimately always lose the fight should not stop them from trying, and it does not—lives are prolonged every day in hospitals and clinics around the world. Realistically, though, some of these medical achievements may amount to no more than a stay of execution. Death is coming, and if it wasn't today, it might be tomorrow.

Over the centuries, society has catalogued and measured life expectancy, by which we mean the age at which we are statistically most likely to die—or, to look at it more positively, the length of time we are likely to spend living. Life tables are interesting and useful tools but they are dangerous, too, in that they create an expectation that will not be reached by some and will be exceeded by others. We have no way of knowing whether we are the average Joe who will conform to the norm or whether we will be an outlier at one end or the other of life's bell curve.

And when we find ourselves to one side or other of the curve, we take it personally. We are proud of ourselves when we exceed our life expectancy because it makes us feel that we have somehow beaten the odds. When we don't reach the age anticipated for us, those we leave behind may feel that they have been robbed of the life of someone dear to them and experience anger, bitterness or frustration. But of course that is simply the nature of the life curve: the norm is just the norm, and most of us will fall into the variations around it. It is unfair to blame death and accuse her of cruelty and larceny when she has always been honest in demonstrating that our life spans can be anywhere within the range of human possibilities.

The longest-living person in the world whose age could be verified was Frenchwoman Jeanne Calment, who was 122 years and 164 days of age when she died in 1997. In 1930, the year of my mother's birth, female life expectancy was sixty-three, so on her death at seventy-seven she exceeded the norm by fourteen years. My grandmother fared even better: when she was born, in 1898, her life expectancy would have been only fifty-two. She lived to be seventy-eight, outstripping that by twenty-six years, which may in part be a reflection of the huge number of medical advances during her lifetime—although her cigarettes didn't help her in the end. The prediction for me, when I arrived in 1961, was a life that might be seventy-four years long. That would leave me now with just seventeen to go. My goodness me, how did that happen so quickly? However, based on my current age and lifestyle, I can now realistically expect to reach eighty-five, and I may have another twenty-nine years or more to look forward to. Phew.

So, during the course of my life, I have gained the prospect of an additional eleven years. Isn't that great? Not really. You see, I didn't get those extra years when I was twenty, or even forty. If I am given them, it will be when I am seventy-four. Would that we could be granted more time in our prime, where youth continues to be wasted on the young.

The calculations of life expectancy given at birth are slowly becoming more accurate and we know that among the next two generations, those of my children and grandchildren, there will be more centenarians than have ever before existed in human history. Yet the maximum age to which our species is capable of living is not increasing. What is changing dramatically is the average age at which we die, and therefore we are seeing an increase in the number of individuals falling into the far right regions of the bell curve. In other words, we are changing the shape of human demography. The rapidly expanding health and social issues created by the growth of an ageing population are starting to give us a glimpse of the resulting impact on society.

While longer lives are for the most part to be celebrated, I do wonder at times if, in striving to stay alive for as long as possible at all costs, all we are in fact doing is prolonging our dying. While life expectancy may be variable, death expectancy remains unchanged. Should we ever actually conquer death, the human race and the planet would be in real trouble.

Working every day with death as my companion, I have come to respect her. She gives me no cause to fear her presence or her work. I think I understand her reasonably well because we choose to communicate in direct, plain and simple language. It is when she has done her job that I am permitted to do mine and, thanks to her, I have enjoyed a long, productive and interesting career.

This book is not a traditional treatise on death. It does not follow the well-trodden path of examining lofty academic theories or quirky cultural variations or offer warm platitudes. Instead I will simply try to explore the many faces of death as I have come to know them, the perspectives she has shown me and the one she will ultimately reveal to me some time in the next thirty years or so, if she chooses to spare me that long. And it is, like forensic anthropology itself, which seeks to reconstruct through death the story of the life lived, as much about life as about death—those inseparable parts of the continuous whole.

In return, I ask only one thing of you: suspend your preconceptions of death for a moment, any sense of distrust, fear and loathing, and perhaps you will begin to see her as I do.

You may even begin to warm to her company, get to know her a little better and cease to be afraid of her. In my experience, engaging with her is both compelling and fascinating, and never dull, but she is complex and sometimes wonderfully unpredictable. You have nothing to lose—and in your own encounters with her, surely it is better to be dealing with the devil you know.

CHAPTER 1

Silent teachers

'Mortui vivos docent'
(The dead teach the living)
Origin unknown

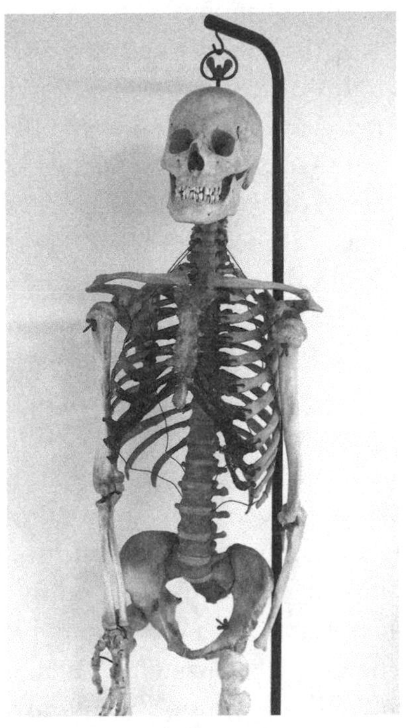

An articulated adult human skeleton which hangs in my laboratory.

From the age of twelve, I spent every Saturday and all my school holidays for five years up to my elbows in muscle, bone, blood and viscera. My parents had a fearsome Presbyterian work ethic and I was expected to get myself a part-time job and start earning some money as soon as I was old enough. So I went to work at the butcher's shop at Balnafettack Farm on the outskirts of Inverness. It was my first and only job as a schoolgirl and I loved every single minute of it. I was utterly oblivious of the fact that most of my friends, who preferred to work in pharmacies, supermarkets or clothes shops, considered it an odd choice, not to say vaguely distasteful. In those days I had no inkling that the world of forensic science was waiting for me but, looking back now, I see this as part of the pattern for my life that was hidden from me, and from them, at the time.

A butcher's shop was an extremely useful training ground for a future anatomist and forensic anthropologist and a happy and fascinating place to work. I loved the clinical precision involved in the butcher's craft. I learned a multitude of skills: how to make mince, how to link sausages and, most importantly, how to make regular cups of tea for the butchers. I learned the value of a sharpened blade as I watched them manoeuvre their knives swiftly and skilfully around irregularly shaped bones, paring away the dark red muscle to reveal the startling clean white skeleton beneath. They always knew exactly where to cut so that the meat could be rolled artistically into brisket or sliced evenly into stewing steak. There was something reassuring about the certainty that the anatomy they encountered would be the same every time. Or almost every time: I do remember the odd occasion when a butcher would curse under his breath about something not being 'quite right'. It seems cows and sheep have their anatomical variations, just like humans.

I learned about tendons and why we cut them out; where, in the space between muscles, there are blood vessels that need to be excised; how to remove the confluence of structures at the hilum of the kidney (too tough to eat) and how to open the joint between two bones to reveal the glassy, viscous fluid of the synovial joint space. I learned that when your hands are cold—and they always seem to be cold in a butcher's shop—you look forward to the delivery of fresh livers, still warm from the abattoir. For a fleeting moment, when you dipped your hands into the box, you could feel them again, thanks to the warm cow's blood de-icing your own.

I learned not to bite my fingernails, never to place a knife on the block with the blade facing upwards and that blunt knives cause more accidents than sharp ones—although sharp blades leave a much more spectacular mess when a mistake is made. I still find it tremendously satisfying to see the neat array of anatomy on display in a butcher's shop, always laid out precisely, cut and prepared the way it should be, and to catch that slight whiff of iron in the air.

I was sad when I had to give up the job. I idolised my biology teacher, Dr Archie Fraser, to such extent that whatever he said I should do, I did it. So when he told me I must go to university, off to university I went. As I had no idea what I ought to study, following in his footsteps and opting for biology seemed a good idea. I spent my first two years at the University of Aberdeen in a bored haze of psychology, chemistry, soil science, zoology (which I failed first time round), general biology, histology and botany. At the end of it all I found I was best at botany and histology, but the pro-spect of studying plants for the rest of my life made my eyes bleed. That left histology, the study of human cells. Having completed the histology module, I felt I never wanted to have to look down a microscope again—everything seemed to consist of amorphous blobs of pink and purple. It was, though, my route into anatomy, where I would be able to dissect a human cadaver. I was only nineteen and had never seen a dead body before, but for a girl who had spent five years of her life cutting up animals in a butcher's shop, how hard could it be?

Perhaps my Saturday job prepared me in a very minor way for what lay ahead. The first experience of a dissecting room is, though, daunting for everyone. It is one of those moments nobody forgets because it assaults every single sense. There were only four of us in the class and I can still hear the echoes that reverberated round that vast, grand room which, with its high, opaque glass windows and intricate Victorian parquet floor, might have served in different circumstances as a conservatory. I can still smell the formalin, a chemical stench so thick you could taste it, and see the heavy glass and metal dissecting tables with their peeling green paint—forty or more of them, set out in regimented rows and shrouded in white sheets. On two of the tables, hidden under their sheets, were the bodies that were waiting for us, one for each pair of students.

It is also an experience that immediately challenges your perceptions of yourself and others. You feel very small and insignificant when it dawns on you that here is someone who, in life, made the choice to give themselves in death to allow others to learn. It is a noble deed that has never lost its poignancy for me. If ever I lose sight of the miracle of that gift, it will be time to hang up my scalpel and do something else.

At random, my dissection partner, Graham, and I had been assigned the cadaver of this selfless donor—a body expertly prepared for us by the anatomy technician that would be our world of investigation for a full academic year. Not knowing his real name, we rather unoriginally called him Henry, after Henry

Gray, author of *Gray's Anatomy*, the text that would come to dominate my life. Henry, a man who hailed from the Aberdeen area and was in his late seventies when he died, had elected to bequeath his body to the anatomy department at the university for the purposes of education and research. My education, and Graham's, as it turned out.

It was sobering to think that at the time Henry had made his decision, I, his future pupil, was completely unaware of the amazingly generous act that would shape my entire life. I would have been busy bemoaning my lot of having to dissect rats in zoology, which I loathed.

When he died, I was probably cutting up another of the university's apparently endless supply of plant stems to study their cellular structure, oblivious of his passing. Every year, when I talk to my first- and second-year students preparing to go into anatomy dissection in their third year, I tell them that the person they will study with, and learn from, is currently still alive. Perhaps that very day someone will be making the decision to bequeath his or her remains for the benefit of their education. I am always reassured when there are a few sharp intakes of breath as the enormity of that concept sinks in. There are inevitably a few who well up at the idea of a person they might have walked past on the street that morning ending up on their dissection table—and so they should. Such a huge gesture by a total stranger should never be taken for granted.

Henry's cause of death was registered as myocardial infarction (heart attack) and his body had been collected from the hospital where he had died and then transported by the funeral director into the care of the anatomy department. Whether he had family, whether they supported him in his decision or how they felt about the lack of the normal ritual of a funeral, I would never know.

In a tiled, dark and clinically soulless room in the basement of the anatomy department at Marischal College, hours after Henry's death, Alec the mortuary technician had removed Henry's clothing and personal effects, shaved his head and attached four brass identification discs—each threaded with a piece of cord and stamped with a sequential identification number—to his smallest fingers and toes. These would stay with Henry throughout his time at the university. Next Alec would have made a cut in the skin of Henry's groin, about 6cm in length, and dissected away the overlying muscle and fat until he could locate the femoral artery and vein in the region of the thigh known as the femoral triangle. He would then have made a small longitudinal incision in the vein, and another in the artery, where he inserted a cannula, securing it in place with some more cord. When a tight seal had been achieved, a valve in the cannula would have been opened and a solution of formalin would have perfused gently through Henry's arborescent arterial system, driven from a gravity-feed tank above him.

The embalming fluid would have found its way via the blood vessels to every single cell in his body—to the neurons in his brain, where he used to think about all the things that mattered to him; to his fingers, which had held the hand of someone he cared about; to his throat, through which his last words had been spoken, perhaps only hours before. As the formalin solution slowly pushed its way onward, in an irreversible wave, the blood in his vessels would have been purged and eventually much of it would have washed away. After only two or three hours of this quiet, peaceful embalming process, his body would have been wrapped in plastic sheeting and stored until it was needed, maybe days, maybe months later.

In that short interval, Henry had been transformed, of his own volition, from a man known and loved by his family into an anonymous cadaver identified only by a number. That anonymity is important. It protects the students and helps them to mentally separate the sad death of a fellow human being from the work they are doing. If they are to dissect a cadaver for the first time without experiencing crippling empathy, they must, while remaining respectful and ensuring that dignity is preserved, be able to train their minds into viewing the body as a depersonalised shell.

When the time came for Henry's body to play its part in our first anatomy class, he had been placed on a trolley, brought upstairs to the dissecting room in the old, rickety, noisy lift, transferred on to one of the glass-covered dissecting tables and covered with a white sheet to wait, quietly and patiently, for his students to arrive.

Today we go to great lengths to make our students' first dissection as memorable and atraumatic as possible. Most of them, like me, will never have seen a dead body before this moment. In 1980, when I embarked on anatomical dissection, there were no introductory sessions, no gradual process of getting to know the cadaver that would be our silent teacher for the next few months. We were four very scared third-year undergraduates who, on arriving that Monday morning armed only with our copies of *Snell's Clinical Anatomy for Medical Students*, a dissection manual—G.J. Romanes' *Cunningham's Manual of Practical Anatomy*—and a selection of scary dissecting instruments wrapped up in a khaki-coloured cloth roll, were pretty much left to just get on with it, beginning at page 1 of the manual. We didn't use gloves or wear eye-protectors, and our laboratory coats very soon became an utter disgrace as we were not allowed to take them out of the building to wash them. How times have changed.

On our table, Graham and I found an array of sponges, which we quickly learned were essential for mopping up excess fluid as the dissection progressed. They had to be wrung out frequently. Underneath it was a stainless-steel bucket for collecting pieces of tissue when our dissection was complete for the day. It is important that all the parts of a body remain together, even when they amount to no more than small bits of muscle or skin, so that when it is sent for burial or cremation it will be as complete as possible. Standing sentinel at our side, watching and waiting, was a second influential tutor: an articulated human skeleton, there to help us understand what we would see and feel under Henry's skin and muscles.

The first thing to master was how to put on a scalpel blade without slicing your finger off. Lining up the narrow slit on the blade with the ridge of the handle, then guiding it with forceps until it clicks into place, takes some dexterity and practice. As does removing it again. I often think to myself that surely someone could come up with a better design.

If you cut into the cadaver and noticed it starting to bleed with bright red arterial blood, I was warned, just remember that cadavers don't bleed. What you will have cut is your own finger. The scalpel blades are so sharp and the room so cold that you don't feel them slicing into your skin. So the first indication that you have injured yourself will be the sight of scarlet living blood pooling against the pale brown of the cadaver's embalmed skin. Contamination is not as much of a concern as it would be if we were handling unembalmed bodies because the process renders the tissue virtually sterile. Just as well, since dealing with fiddly little blades when your fingers are cold and slippery with body fat is not easy. These days we begin the academic session with a vast supply of sticking plasters and surgical gloves.

Once the blade is finally on your scalpel handle and your finger has stopped pumping blood, you lean over the table and immediately your eyes start to water from the formalin fumes. The manual has told you where to cut, but it doesn't tell you how deeply, or what it will feel like. Nobody has given you explicit permission to 'feel' Henry's anatomy so that you can figure out where to cut to and from, and none of it seems to make any sense. It is all a bit terrifying and faintly embarrassing. You pause for a moment to consider how you will make the incision down the centre of the torso, from the hollow of the sternal notch at the base of the neck to the lower border of the ribcage. Which of you will watch and which of you will make the cut? Your hand shakes. That first incision stays with every student, however blasé they pretend to be. If I close my eyes I can still remember what it was like, and how impeccably Henry tolerated our youthful ineptitude.

As your motionless teacher reclines in patient repose, waiting for you to begin, you inwardly apologise to him for what you are about to do, for fear you will make a mess of it. Scalpel in the right hand, forceps in the left . . . how deep do you cut? It is no coincidence that most students begin dissection with the thorax. The breastbone is so close to the skin that no matter how hard you try, there is little you can do wrong. You simply cannot go too deep. You lower the blade to the surface of the skin and draw it carefully down the chest wall, leaving a faint line in its wake.

It is surprising just how easily the skin parts. It is leathery to the touch, cold and wet, and as it separates from the tissue, beneath the blade you glimpse the contrasting pale yellow of the subcutaneous fat. Feeling a little more confident, you extend the incision from the sternum in the centre across both clavicles, out towards the tip of each shoulder and you have made your first 'T' postmortem cut. So much anxious anticipation and it is over in a moment. The world has not stopped. The relief is immense and only now do you realise that you have not breathed through the whole process. Though your heart is racing and your adrenaline is pumping, you are surprised to find that you are no longer afraid but intrigued.

Now you need to expose the tissue underneath. You start to peel back the skin, picking carefully at the corner of the free flap in the midline above the breastbone, at the junction of the two limbs of the 'T'. You grip the skin with the forceps, applying just enough tension to allow the blade to separate it from the tissue. You never really need to cut. The yellow fat appears and as this comes into contact with your warmer hands it liquefies. Holding the scalpel and forceps suddenly becomes tricky and the flicker of confidence you had a few moments earlier evaporates as the forceps slip off the skin and fat and fluid splash up into your face. Nobody has warned you about this. Formalin smells nasty but it tastes worse. You only ever make that mistake once.

Continuing to peel back the skin, you start to notice tiny red dots and realise that you have unavoidably cut across a small cutaneous blood vessel. Suddenly, the immense scale of the human form, and the vast amount of information it contains, hits home. The day before you might have been wondering how on earth it was going to take you a whole year to dissect a human body and why you needed three whole dissector volumes to instruct you. Now it dawns on you that a year will be nowhere near long enough to do much more than scratch the surface of your subject. You feel like the true novice you are. You despair that you will ever remember all you must learn, let alone understand it fully.

You put a little strain on the forceps and the sharp blade slices into the connective tissue with disconcerting ease, even though it seems to you that your scalpel is barely touching it. As the underlying muscles are revealed, the white transverse bony walls of the chest stand out starkly against them like a bleached toast rack. Your eyes trace the shape of the hollows and ridges of the skeleton at your side as you feel beneath your fingertips Henry's muscle and bones. You begin to identify and name the bones and their constituent parts—the scaffolding of the human body—and before you know it, you are speaking an ancient language understood by anatomists all over the world: a language that would have been familiar to Andreas Vesalius, the fourteenth-century founder of modern anatomical study and my undisputed girl crush.

At first, the embalmed muscle appears to be a uniform light brown mass (disarmingly, slightly reminiscent of tinned tuna fish), but as you look closer, and your eye starts to attune to its patterns, you can make out the orientation of its fibres and the thin strings of the nerves that supply it. You locate the origins and the insertions of the muscle and deduce its action on the joint it crosses, captivated now by the wonderfully logical engineering you are examining. As a living person, you remain separate from death, but the mesmerising beauty of human anatomy has created a bridge into the world of the dead, one that few will cross and none who do will ever forget. The sensation of traversing that bridge for the first time is an experience you can never repeat. It is special.

The study of anatomy polarises its students: they either love it or they hate it. The fascination lies in the logic and order of the subject; the downside is the vast amount of information to be learned—that and the smell of formalin. When the fascination outweighs the drawbacks, anatomy imprints itself on your soul and you will consider yourself for ever a member of a privileged elite: the select few who have seen and been taught the secrets of human construction by those who have chosen to allow you to look inside their own bodies. We may stand on the shoulders of scholarly giants, Hippocrates and Galen, and their descendants Leonardo Da Vinci and Vesalius, but the real heroes are undisputedly those extraordinary men and women who choose to bequeath their mortal remains so that others may learn: the anatomy donors.

◊

Anatomy teaches you many things beyond the workings of the corporeal form. It teaches you about life and death, humanity and altruism, respect and dignity; about teamwork, the importance of attention to detail, patience, calmness and manual dexterity. Our interaction with the human body is tactile and very, very personal. No book, model or computer graphic will ever come close to dissection as a means of learning our craft. It is the only way to do it if you are to become a card-carrying anatomist.

It is, though, a subject that has been much maligned as well as revered in its past. Since the glorious years of the early anatomists, from Galen to Gray, right up to the present day, it has been tainted periodically by nefarious characters who have sought to exploit it for profit. In nineteenth-century Edinburgh, the heinous deeds of Burke and Hare, who turned to murder to supply cadavers to anatomy schools, led to the passing of the Anatomy Act of 1832. As recently as 1998, the sculptor Anthony-Noel Kelly was jailed for stealing body parts from the Royal College of Surgeons in a case that cast a spotlight on the ethics of art and the legal status of human remains donated to medical science. And in 2005, an American medical tissue company was closed down after its president was convicted of illegally harvesting body parts and selling them on to medical organisations. It seems that anatomy is not immune to the economics of supply and demand, or to the criminal acts of a few racketeers with no regard for decency, dignity or decorum. It is right therefore that we defend our donors and that they are protected by an Act of Parliament.

There is money in death, and where there is money to be made there will always be those prepared to cross ethical boundaries to make more of it. Given that the sale of human remains is legal in many countries and that a good number of institutions around the world will pay a hefty price even for an articulated human skeleton, perhaps it should not surprise us that the ancient crime of grave-robbing persists in modern-day forms. When I was a student in the 1980s, most of the teaching skeletons used in dissecting rooms were imported from India, long regarded as the world's primary source of medical bones. Although the Indian government outlawed the export of human remains in 1985, a global black market still thrives there today. In the UK, we have rightly become intolerant of the sale of bones or any other body parts.

What is or is not considered acceptable in terms of the treatment of human remains fluctuates, like all social attitudes, and can sometimes change quite markedly in the course of a single lifetime. The skeletons currently used to teach UK anatomy students are more likely to be plastic replicas, and although human ones can still be found in the dusty old cupboards of school science labs, GPs' surgeries and first-aid training facilities, many organisations that possess them quite legally are uncomfortable nowadays about holding on to them. Some opt to donate them to a local anatomy department and in return they might be offered an artificial teaching skeleton as a replacement.

Unlike our predecessors, contemporary anatomists can take time over dissection and thus gain far more value from our cadavers in our study of the infinitesimal detail of the human form, thanks mainly to centuries of research into ways of preserving the human body and halting the process of decay. Since the early days of dissecting corpses freshly cut down from the gibbet, anatomists have striven to preserve cadavers for as long as possible by following the techniques developed by the food industry, learning how to pickle in alcohol or brine or how to desiccate and freeze.

After Lord Nelson's death in the Battle of Trafalgar in 1805, his body was stored in a vat of 'spirit of wine' (brandy and ethanol) for his journey home to a hero's funeral. Pickling alcohol remained the preferred method of preservation until the discovery of a nasty chemical called formaldehyde later in the nineteenth century went on to transform the field of anatomy. Formaldehyde is a disinfectant, a biocide and a tissue fixative and works so well that its aqueous solution, formalin, is still the most commonly used preservative worldwide.

But in sufficient concentration formaldehyde is a hazard

to human health and recent decades have seen alternatives being considered. These include fresh-frozen cadavers, where the body is dismembered into parts that are frozen and then thawed as required for dissection, and soft-fix methods that leave the body more supple and closer in texture to a living human. In the 1970s, the anatomist Gunther von Hagens pioneered plastination, whereby water and fat are removed under vacuum and replaced by polymers. These body parts have eternal life. As they will never decompose, we have succeeded in designing a new environmental pollutant.

Whatever advances there may be in the technology we use to preserve bodies or to investigate them through medical imaging, anatomy itself, of course, does not change. What was seen in the cadavers dissected by Vesalius in 1540 or by Robert Knox in 1830 was, in essence, no different from what Graham and I saw during the academic year we spent with Henry. However, as Vesalius and Knox were obliged to dissect fresh remains, the very limited time they had with a cadaver probably did not engender the same bond of trust and respect between the dissector and the dissected that I was fortunate enough to be able to establish with Henry. Or maybe it is just that social and cultural attitudes have changed down the years.

For me there can never be another Henry, and for every anatomist their own Henry will be special. I learned so much in that year about myself as well as about the human form. At those stages in our lives when we look back to pinpoint the times that made us happy and fulfilled, my search will always lead back to Henry. There are few moments in that year that I would trade, but I would be lying if I said there were none. I hated cutting through the beds of his fingernails and toenails as I always felt, irrationally, that it would hurt. And to be honest, nobody enjoys flushing out the digestive system.

But for me the rewards to be gained from the study of the dead far outstrip those less palatable moments, and the gut-wrenching fear that kicks in when you become properly aware of the sheer volume of what you need to master: over 650 muscles must be committed to memory, along with their sites of origin and insertion, their nerve supply and their actions; more than 220 named nerves, their root values and whether they are autonomic, cranial, spinal, sensory or motor; hundreds of named arteries and veins that spread in an arborescent pattern from the heart and back again, their origins, their divisions and the related soft-tissue structures. Then there are the 360-plus joints, and don't get me going on the three-dimensional relations of the developing gut, tissue embryology, neuroanatomy and its tracts.

Just when you think you are getting to grips with some of these anatomical structures, they skip from your fingers like soap in the shower and you have to start all over again. Utterly infuriating. But the reiteration of mountains of facts and connections is the only way to learn and understand the three-dimensional complexity of the human. Anatomists don't have to be particularly clever: they just need a good memory, a logical learning plan and spatial awareness.

Henry allowed me to probe into every detail of the workings of his body, to explore his anatomical variations (bless him and his aberrant superficial epigastric artery—I will never forget that!), to get frustrated when I cut through something when I should have known better and helped me to wrestle with the virtually invisible parasympathetic nervous system. He bore it all with fortitude, never scolding me or making me feel foolish, and in time the balance tipped to a point where I was learning more about him, in one sense, than he might ever have known about himself.

I discovered that he didn't smoke (his lungs were clear), he didn't drink to excess (his liver was in very good condition), he was well nourished but didn't overeat (he was tall and slim, with little body fat, but not emaciated), his kidneys looked healthy, his brain showed no tumours and there was no evidence of aneurysm or ischaemia. While his cause of death was listed as a myocardial infarction, his heart looked strong to me. But what did I know? I was just a third-year rookie.

Perhaps he died simply because it was his time to die, and something likely had to go down on his death certificate. The cause of death given for a cadaver often elicits concern in our students when, on coming to examine the organ in question, they find no disease or abnormality. When death is due simply to old age and it is known that the deceased desired to bequeath their body, the recorded cause of death will inevitably be reasoned and educated supposition. The only way to establish it for certain would be to perform a postmortem which, as this procedure renders the body useless for dissection, would contravene the wishes of the donor. So as long as the death is not suspicious and is consistent with the age of the deceased, in many cadavers the cause will have been rationally deduced as heart attack, stroke or pneumonia—dubbed by some the old person's friend.

By the time we had finished cataloguing Henry's body, all the way from the top of his head to the tip of his little toe, there was no part of him we had left unexamined. No part that we had not pored over in books, debated, checked and confirmed. I was so proud of this man I would never know as a living, breathing, talking, active person but with whom I was now so personally and intimately familiar that I felt I knew him in a way nobody else ever had, or ever would. What he taught me has stayed with me, and will stay with me, for ever.

Within a few months, it would be time to say goodbye and to promise Henry that I would put the education he had given me to good use. I bid him a final farewell at King's College chapel in Aberdeen at a moving service of thanksgiving for the gift of our bequeathers, attended by their families and friends, staff and students. I couldn't know, when the names were read out, which one was Henry's. From my hard, wooden seat in the choir stall, I scanned the faces of the congregation, wondering which of the grieving relatives was shedding a tear for him. Which of the people sitting on those well-worn pews had been his *amicus mortis*, his friend in death? I so hoped he hadn't died alone. It is more comforting to think that someone dear to him had been at his side to hold his hand and tell him he was loved.

All of the Scottish anatomy schools arrange these services every year. They allow us to pay our respects and demonstrate to the families and friends of the bequeathed just how crucial their gift has been, how much it was valued and how important in fostering the education of the next generation.

CHAPTER 2

Our cells and ourselves

'Without systematic attention to death, life sciences would not be complete'

Elie Metchnikoff, microbiologist (1845–1916)

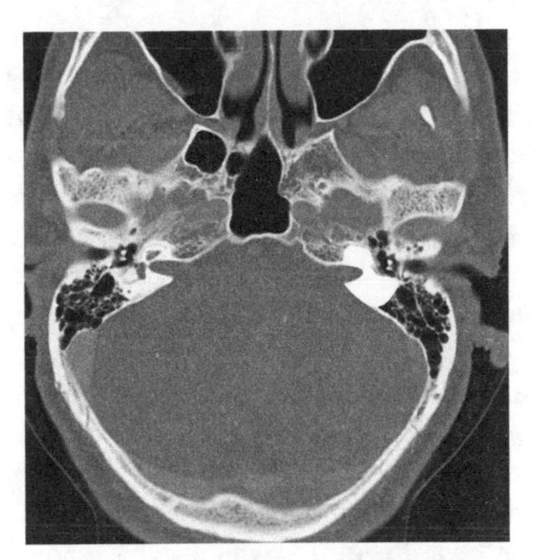

A CT scan of the skull showing the position of the otic capsule at its base.

What makes us human? One of my favourite definitions is: 'Humans belong to the group of conscious beings that are carbon-based, solar system-dependent, limited in knowledge, prone to error and mortal.'

It is strangely comforting to be granted tacit permission to make mistakes just because we are human. As we possess neither the capability to get everything right first time nor an unlimited lifespan at our disposal to practise and hone each task to perfection, we should accept that our lives will be a bit of a mixture. Some tasks we will fulfil well and they will enrich our lives and the lives of others; those we will clearly never master are just a waste of our precious time.

There is a lovely moment in the film *When Peggy Sue Got Married* which epitomises the human desire for glimpses into the future to help us focus our attention today on what will ultimately prove worthwhile—or not. 'I happen to know that in the future,' Peggy Sue tells her teacher after a maths test, 'I will not have the slightest use for algebra—and I speak from experience.' Forward planning when we have no idea what lies ahead of us is tricky and, while it can seem unimportant when we are young, as we near our allotted three score years and ten, life seems to speed up and we begin to become aware of how much we still have left to achieve.

The 'conscious' aspect of being human is perhaps our most defining characteristic. This centres on our knowledge of 'self'—the almost unique ability to display introspection, and thereby to recognise ourselves as separate individuals from others. The psychology surrounding identity and recognition of 'self' is extremely complex. In the 1950s, the developmental psychologist Erik Erikson summarised identity as: 'Either a) a social category, defined by membership rules and (alleged) characteristic attributes or expected behaviours, or b) socially distinguishing features that a person takes special pride in or views as unchangeable but socially consequential (or a) and b) at once).'

Researchers believe that a sense of identity is a manifestation and extension of the maturation of the concept of self which allows us to develop an intimate and intricate society. It enables us, to a certain degree, to express individuality, and perhaps helps others to tolerate it, by permitting us to both promote and display who we are, who we want to be and what we choose to stand for. Thus we can actively draw like-minded people around us and repel those with whom we do not, or do not wish to, identify. This freedom of individuality, and indeed its suppression, gives humans a unique capability and opportunity to play with their identity and to manipulate, or even change, the perception, portrayal and concept of 'self'. It is here that I think Erikson omitted the third and most important category of identity, and the one that is most fun to play around with: physical identity.

If, as a species, we recognise the physical differences between self and others, then we can use this ability to try to differentiate between any two individuals. The importance of identity in our society, and the fact that it can be manipulated, places it at the core of investigative sciences, including my world of forensic anthropology—the identification of the human, or what remains of the human, for medico-legal purposes.

How can it be proved, using our innate human biology or chemistry, that we are who we say we are, and that who we say we are is who we have always been? Forensic science can be used as a toolbox of techniques to reconcile an unidentified body with its previous living identity. Forensic anthropologists look to features of our corporeal biology or chemistry to analyse a trackable and readable history of the life lived, and to confirm whether the evidence recovered matches traces left by that person in the past. In other words, we search for clues of the narrative written in our bodies, innate and acquired, laid down between birth and death.

From a much more pedestrian biological perspective, the human can be rather crudely defined as a large mass of self-regulating cells. If histology, the study of the microscopic anatomy of the cells and tissues of plants and animals, and the cell cycle have never got me terribly excited—there is just far too much complex biochemistry involved for my simple little brain to compute or to be bothered with—we must acknowledge that the cell is the basic unit of all known living organisms. So if death is going to be held responsible for the end of an organism's existence, then she is also going to have to take the rap for the death of every cell. Anatomists know that ultimate organismal death can often be traced from the cell to the tissue and then to an organ or an organ system. So, whether we like it or not, it all begins and ends with cells. Death may be a single event for the individual but it is a process for the body's cells, and to understand how that works, we must be familiar with the life cycle of the building blocks of the organism. Stay with me—I promise it won't be too boring . . .

Every human is created when two separate cells fuse and then begin to multiply—an incredibly humble beginning from an unimpressive little sack of proteins. After forty weeks *in utero*, those two cells will have gone through the most miraculous transformation, becoming a highly organised mass of over 26 billion. The huge increase in the size of the fetus and in the specialisation of its individual components requires a tremendous amount of precise planning if everything is to go as it should, and, thankfully, much of the time it does. By the time that baby becomes an adult, the cell mass will have expanded to over 50 trillion, grouped into some 250 different cell types forming four basic tissues—epithelial, connective, muscular and nervous—and a variety of sub-tissues. These in turn will combine to construct approximately seventy-eight different organs, divided into thirteen major organ systems and seven regional groupings. Remarkably, only five organs are considered vital to sustained life: the heart, brain, lungs, kidneys and liver.

Every single minute about 300 million of our cells will die, many of which are simply replaced. Our bodies are programmed to know which cells to replace, and when and how, and by and large they just get on with it. Each cell, tissue type and organ has its own life expectancy, which is managed like stock turnover in a supermarket based on a 'best before' date. Somewhat ironically, those with the shortest shelf life are the ones that start it all off: sperm survive for only three to five days after formation. Skin cells live for a mere two to three weeks and red blood cells only three or four months. Not surprisingly, there is greater longevity in the tissues and the organs. The liver takes a full year to replace all its cells and the skeleton almost fifteen years.

The charming myth that, because we replace so many of our cells on a regular basis, every decade or so we become an entirely new physical person is, sadly, misguided. No doubt it has its roots in the Theseus paradox—the question of whether an object that has had all of its components replaced at some stage remains fundamentally the same object. Can you imagine how we might be able to play with this fantastical concept in a court of law? Picture a wily old defence barrister holding forth in a murder trial. 'But Your Honour, my client's wife died fifteen years ago, so even if the person he once was killed her, he is no longer physically the same person because every cell in his body has now died and been replaced. The man before you could not have been at the crime scene because he did not exist.'

I don't believe such an argument has yet been rehearsed in court, but I would dearly love to be the Crown witness if ever it is. It would be fun to engage in these metaphysical musings with a barrister. It does, though, raise a question: just how much alteration can a biological entity sustain while remaining recognisable as the same individual and maintaining its traceable identity? Take the physical changes witnessed in the late Michael Jackson over the years. Much of the child star of the Jackson Five was almost unrecognisable in the adult he became, but there would have been other components that persisted throughout his life and which continued to anchor his physical identity. Our job is to find such components.

There are at least four cell types in our bodies that are never replaced and which can live to be as old as we are—technically even longer, in the case of those formed before we are born. Perhaps these cells might be cited as the unlikely seat of our corporeal biological constancy to confound that tricky defence lawyer's argument. The four permanent cell types are the neurons in our nervous system, a tiny little area of bone at the base of our skull called the otic capsule, the enamel in our teeth and the lenses in our eyes. Teeth and lenses are only semi-permanent as they can be removed and substituted by modern dentistry or surgery respectively without harming the host. The other two are immovable and therefore truly permanent, remaining locked in our bodies as irrefutable evidence of our biological identity from before our birth until after our death.

Our neurons, or nerve cells, are formed in the very early months of embryonic development and by the time we are born we will have as many as we are going to have for the rest of our lives. Their axons, which resemble long, extending arms, branch out like a motorway system, conveying traffic from north to south and vice versa. They carry motor commands south from the brain to our muscles and sensory information in the opposite direction from our skin and other receptors. The longest are those that transmit pain and other sensations along the full length of the body, from the tip of the little toe, all the way up through the foot, leg and thigh, up the spine and the brain stem and on to the sensory cortex of the brain at the top of the head. If you are six feet tall, each single neuron on that pathway may be close to seven feet long. So when we stub our little toe on the wardrobe, the message takes a moment to reach our brains, which is why we may have a pain-free split second before we feel the 'ouch' we know is coming.

It is the persistence of these cells within the brain that gives rise to the interesting question of whether there is an aspect of our identity to be found there. It is possible that the pattern of communication between them could be mapped to show how we think and how the higher functions of reasoning and memory come about. Recent research has demonstrated that with the help of a fluorescent protein we can now see a memory being formed at the single synapse level. Practical application may yet be a little too much science fiction for us to embrace fully, although I am tempted to predict that an understanding of the key role neurons may play in establishing identity might not be so very far away.

The second seat of cellular permanency is in the otic capsule, situated in the very depths of the skull around the inner ear. This is part of the petrous temporal bone, which houses the cochlea, the organ of hearing, and the semi-circular canals responsible for balance. As the inner ear forms in the embryo and fetus, it does so to full adult size immediately and remains insulated against growth and remodelling through the production of high levels of osteoprotegerin (OPG), a basic glycoprotein that suppresses bone turnover. It does not, in normal circumstances, remodel because if it were permitted to grow, it would interfere with the intricacies of our hearing and balance. Even though the otic region is already adult-sized in newborn babies, it is very, very small, representing, in volumetric terms, only around 200 microlitres—about the size of four raindrops. Unlike the neurons, the cells locked in this little bone already offer us opportunities to recover information about individual identity.

To understand what value any cell may have in the process of human identification we need to know how they form—be they bone cells, muscle cells or those that line the gut. At its most basic level, every cell in our body is comprised of chemicals. Their formation, survival and replication are dependent on a supply of elemental building blocks, an energy source to bind them together and keep them alive and a waste-disposal outlet for their by-products. The main opening in our bodies by which the building blocks for future cellular construction can enter is the mouth, leading on to the stomach and the gut system—our food-processing factory. So the core components of every single cell, tissue and organ can be obtained only from what we ingest. We are, literally, what we eat. Refuelling, then, is vital to survival, and the maxim that we cannot survive without air for more than three minutes, water for more than three days and food for more than three weeks, while not entirely accurate, is pretty close to the truth.

In utero, before we can ingest food independently, we source our fuel from our mother's diet via the placenta and umbilical cord to enable us to go about the business of developing and organising our own cellular construction. While it is a fallacy that a pregnant woman eats for two, she does need to ensure that her diet is sufficient to meet not only her own needs but also those of a very demanding passenger.

The nutrient building blocks required to construct our otic capsule were supplied by Mum from what she was eating around sixteen weeks into her pregnancy. So within our head, in that minute piece of bone just big enough to hold four raindrops, we will perhaps carry for the rest of our lives the elemental signature of what our mother had for lunch when she was four months pregnant. Proof, if any were needed, that our mums never leave us, and a whole new perspective on the mystery of how they manage to get inside our heads.

We believe we have a cosmopolitan diet but in reality, much of our water and food intake is very local to where we live. As water percolates through various geological formations, it will take up isotope ratios of elements specific to that location and when we ingest it, its signature will be transferred into the chemical make-up of all our tissues.

The chemical composition of our tooth enamel remains largely unchanged throughout life, which is why decaying teeth cannot repair themselves. The crowns of all our deciduous teeth (milk or baby teeth) are formed before we are born and their composition is therefore also directly associated with maternal diet, as is that of the crown of our first adult molar. The rest of our permanent teeth are made by us and reflect our diet during childhood.

As well as our 'permanent' tissues, hair and nails are rich sources of information about diet as their structure is laid down in a linear fashion and they grow at a relatively regular rate. They provide a potential chemical timeline for the deposition of metabolised ingested nutrients that can be read almost like a barcode.

So how can forensic anthropologists use this amazing information offered by our cells to unlock a part of an individual's life story and help confirm their identity? Stable isotope analysis is a good example of one of the scientific techniques that can assist us. The ratio of carbon- and nitrogen-stable isotopes in our tissues may tell us something about diet: whether a person was a carnivore, a pescatarian or a vegetarian. The oxygen isotope ratios may reveal more about the source of water in the diet, and from the stable isotope signature associated with water, we may be able to deduce where they have been living.

If you move to another geographical area, the signature you lay down will alter because of changes in the chemical composition of the food you eat and the water you drink. Analysis of hair and nails can produce a sequential timeline for geographical relocation. This can be extremely useful in trying to identify an unknown deceased person, or to track the movements of criminals. A terrorist suspect, say, who is falsely insisting he has never left the UK may be found out by a sudden change in hair stable isotope ratios that now conforms to a signature you'd expect to find in Afghanistan. Hair analysis can also tell us about persistent consumption of a variety of substances, including drugs such as heroin, cocaine and methamphetamine. And it was, of course, the favoured method of proving arsenic poisoning in Victorian murder mysteries.

So we could, in theory, look at the remains of an individual and, from the isotopic signatures in the otic capsule and first molar, discover where in the world their mother was living when she was pregnant with them and the nature of her diet. We could then analyse the remainder of the adult teeth to establish where the deceased person had grown up, and then the rest of their bones to determine where they had lived for the past fifteen years or so. Finally, we could use their hair and nails to locate where they spent the last years or months of their life.

The complexity of managing the human cell mass is staggering. As a cell factory, the body works incredibly smoothly most of the time when we are at the peak of our fitness, efficiently replacing the majority of those 300 million cells we lose every minute. But as we grow older and degeneration sets in, we become less able to produce new cells. Signs of ageing start to appear: hair becomes thinner and loses its colour, eyesight fades, skin wrinkles and stretches, muscle mass and tone are lost, memory and fertility decline.

These are all evidence of a normal slowing process, or senescence, and clear indicators that we are now probably nearer to the end of our life than its beginning. Being told by your doctor that a condition you have is normal for your age is of little comfort when you realise that death is also normal for your age. To compound the problem of ageing, some cells may 'go rogue' and start to grow and replicate abnormally, tissues damaged by environmental toxins or an abusive lifestyle may cease to function and organs under stress may stop operating effectively. We can extend the longevity of many of our body functions through medical or surgical intervention and pharmacological support, but in the end, when they cannot continue unassisted, we, the organism, will die.

According to one medico-legal definition, organismal death occurs when 'the individual has sustained either irreversible cessation of circulatory and respiratory functions or irreversible cessation of all functions of the entire brain, including the brain stem'. The word 'irreversible' is key. Reversing the irreversible is viewed by the medical community as the holy grail of combating death.

It seems that the activities of those five vital organs define our life and therefore perhaps ultimately our death. The wonders of modern medicine make it possible for us to transplant four of them: heart, lungs, liver and kidneys. But the 'big one', the brain—the fundamental command control centre for every other organ, tissue and cell in our body—has never been successfully replaced. The pact between life and death seems to lie in those neurons (I told you they were special).

◊

Our bodies change not only throughout our life, but also in death. As the processes associated with organismal and cellular deconstruction begin, so we start to break down into the chemical components that were used to build us in the first place. There is an army of volunteers waiting to lend a hand, including the 100 trillion bacteria within the human biome, no longer restrained by an active immune system. Once the dynamic of the environment shifts catastrophically against the likelihood of successful revival or resuscitation of the organism, the bacteria can take over. Death is now confirmed by the fact that life cannot come back.

In most cases, for example, when we die at home with our family around us, or in hospital, or with emergency services in attendance, recording a reliable time of death is relatively straight-forward. However, when someone dies alone, or a body is discovered unexpectedly in possibly suspicious circumstances, we must estimate the time and date of death to fulfil the legal and medical requirements. We try to establish a time death interval (TDI) from information that the body releases. So forensic anthropologists need to understand not only how the body is built, but how it deconstructs.

There are seven recognisable stages of postmortem alteration. The first is pallor mortis (literally, 'paleness of death'), which starts within minutes and remains visible for about an hour. It's what we are referring to when we say that someone who isn't well looks like 'death warmed up'. As the heart ceases to beat, capillary action is halted and blood drains away from the skin surface and begins to settle at the lowest gravitational level within the body. As it happens so early in the postmortem process, this pallor is of very little value in establishing a TDI. It is also a subjective feature and therefore difficult to quantify with any degree of certainty.

The second stage, algor mortis ('coldness of death', or the death chill) follows quickly as the body starts to cool down (in some instances it may warm up, depending on the ambient temperature). The temperature of the body is best recorded from a rectal reading as the skin surface will generally cool, or heat up, more quickly than the deeper tissues. Although the rate of cooling recorded for rectal temperature is relatively constant, it cannot be assumed that the temperature of the body was normal at death. Many factors can influence core temperature, including age, weight, illnesses and medication. Certain infections or drug reactions can raise it, as can exercise or extensive struggling before death; lower than normal readings may be attributable to other physical states such as deep sleep. So this is not an infallible indicator of the TDI.

When someone is found dead, the environment in which their body is discovered will also affect the rate of body cooling. For example, in a location warmer than 37°C a body will not cool and so a calculated TDI based on temperature will be irrelevant. Obviously, if a person has been dead for some time, assessing algor mortis will also be irrelevant as the body temperature will eventually adjust to the ambient temperature.

Within a few hours of death, muscles start to contract and the third and temporary postmortem condition, rigor mortis ('stiffness of death'), will set in. Rigor starts in the smaller muscles first, usually inside five hours, and then spreads to the larger muscles, peaking between twelve and twenty-four hours postmortem. When we die, the pump mechanism that keeps calcium ions out of the muscle cells ceases to operate and calcium floods across the cell membranes. This causes the actin and myosin filaments within muscle to contract and then lock, shortening the muscle. Because muscles cross joints, the joints may also contract and fix into a rigid position in the hours immediately after death. In due course, the rigid muscles will begin to relax due to natural decomposition and chemical alteration and so the joints, too, will become moveable. This is the explanation for the rare but recorded phenomenon of a dead body appearing to twitch or move. But I promise you, the dead do not sit up and moan—that really is something that only happens in horror movies.

Signs of initial flaccidity, rigor and then secondary flaccidity can be used to assist with establishing a TDI but many variables will affect how long rigor lasts and indeed whether it occurs at all. For example, it is quite common for newborn babies and the elderly not to exhibit rigor. In higher temperatures onset is more rapid; lower temperatures delay it. Other factors that will have a bearing include some poisons (strychnine hastens the rigor process, whereas carbon monoxide slows it down). Rigor is also brought about more quickly by intense physical activity before death and may never take place in cases of cold-water drowning. So again, it is not an incontrovertible indicator of TDI, regardless of what may be said in television crime dramas.

With the heart no longer pumping, the body enters the fourth phase of postmortem alteration, livor mortis ('blue colour of death'). The blood will have begun to pool at the lowest gravitational level of the body almost immediately after death, at the pallor mortis stage, but lividity may not be visible for a couple of hours.

The heavier red blood cells sink through the serum and accumulate in the lower gravitational regions. Eventually the skin here will turn a darker red or purple-blue colour as a result of the concentration of red blood cells, in striking contrast to the pallor of the skin at the higher levels. Where the skin comes into contact with a surface (for example, if the body is lying on its back), blood is pushed out of the tissues into adjacent areas where there is no contact pressure. So the contact areas will appear paler compared to the darker surrounding areas of lividity.

Maximum lividity is generally reached within twelve hours. The livor colouration then becomes fixed and can be a useful indicator in the investigation of suspicious deaths. It may reveal how a body was positioned in the hours immediately after death and help us to evaluate whether it might subsequently have been moved. A body with marks of lividity on its back, but found lying on its face, has clearly been turned over. If someone has died by hanging there will be a pooling of blood in the lower segments of all four limbs and this fixed-lividity pattern will remain in the distal arms and legs even after the body is cut down.

A relatively new field of research has been reported in the last year or so concerning the necrobiome—the colonies of bacteria that flourish on the dead. Researchers sampling the bacteria from the ears and nasal openings of cadavers have found that, using next-generation metagenomic DNA sequencing, they may be able to predict TDI very accurately, perhaps to within a couple of hours, even where death has occurred days or weeks before. If this research withstands scrutiny, and the methodology isn't too expensive to run, this new kid on the block may eventually displace the brothers Pallor, Algor, Rigor and Livor.

If a body remains undiscovered through these four phases it will begin to smell pretty bad. In the fifth or putrefactive stage, the cells start to lose their structural integrity and their membranes begin to break down due to the slight acidity of the body fluids. This is called autolysis (literally, 'self-destruction') and provides the perfect conditions for anaerobic bacteria to multiply and start to consume the cells and tissues. This process releases a variety of chemicals, including propionic acid, lactic acid, methane and ammonia, whose presence can be used to detect where a decomposing body has been hidden or buried. We're all familiar with how cadaver dogs are used to search for bodies. Their noses are said to be 1,000 times more sensitive than those of humans and they can sniff out minute quantities of putrefactive scents. Dogs are not the only species with a highly developed sense of smell: rats have also been trained to respond to the odours of decomposition as, incredibly, have wasps.

With the increase in gas production, the corpse will start to bloat and—as some of the odiferous substances, such as cadaverine, skatole and putrescine, grow more concentrated—become irresistible to insect life. Blowflies, in particular, will have detected the putrefaction products within minutes of death, and begun searching for areas in which to lay their eggs, or oviposit, usually around orifices such as the eyes, nose and ears. The unmistakable stench of putrefaction is all-pervasive and insects identify it as a food source for both themselves and their future offspring. The continued build-up of pressure from within the putrefying tissues will lead to purging of liquid at the orifices and can occur to such an extent that the skin splits, allowing even greater access to insects and scavengers. The skin starts to change colour, turning a deep purple, black or an unpleasant green reminiscent of very heavy bruising, due to the decomposition of the by-products of haemoglobin degeneration.

Active and advanced decay, the sixth stage of decomposition, will be set in motion when the larvae hatch and maggot masses take full hold. They will begin in earnest to break down the tissues that have become their food source. Through phases of advanced decay and successive waves of insect, animal and plant activity, all the soft tissue will eventually be consumed. This stage sees the greatest mass tissue loss as a result of feeding and liquefaction into the surrounding environment. The process generates a huge amount of heat: a maggot mass of around 2,500 can raise temperatures to some 14°C above the ambient temperature. Beyond 50°C the larvae will not survive, so when the core of the maggot mass approaches this critical temperature, they will separate and break into smaller and smaller clumps to try to cool down. It is this constant movement away from the central core and the frenzy of insect activity that gives rise to the apt phrase 'a boiling mass of maggots'.

The seventh and final stage is skeletonisation, where all

the soft tissues of the body are lost, leaving only the bones, and possibly some hair and nails, which are made of inert keratin. Depending upon the environmental conditions, and with the passage of enough time, even the bones may be destroyed. We all, then, return to the elements from which we were formed at the start of our life. The planet has finite mineral resources and each of us is made up of recycled parts that we, in turn, give back to the chemical pool.

So how long does this after-death disintegration process take to complete? There is no simple answer. In parts of Africa, where insect activity is voracious and temperatures high, a human body can be rendered from corpse to skeleton within seven days. However, in the cold wilds of Scotland it might take five years or more. As the rate of corporeal decomposition will be influenced by climate, availability of oxygen, cause of death, burial environment, insect infestation, destruction by scavengers, rainfall and clothing, among many other factors, it is not surprising that determining the TDI can rarely be definitive.

The fact that decomposition may be significantly delayed, or indeed halted, either by accident or intent, can also affect the reliability of assessing TDI. Freezing can stop decomposition almost completely and as long as the body doesn't thaw out too many times, recognisable features may remain for centuries. At the other extreme, dry heat, which dehydrates the tissues, can also preserve a corpse. These conditions account for the longevity of, for example, the mummies of Xinjiang and those found in the Spirit Cave in Fallon, Nevada. Chemicals are largely responsible for the prolonged preservation of the famous mummies of Egypt such as Rameses and Tutankhamun. Here, removal of the internal organs and the packing of the body cavities with herbs, spices, oils, resins and natural salts, such as natron, were highly skilled procedures.

Submersion in water, as in the case of bodies found preserved in peat bogs, can halt aerobic activity. The body becomes sterile, and in time the acidic nature of the peat dissolves away the skeleton, leaving behind the tanned leather skin, which may remain almost visually recognisable, even after centuries have passed. In the right conditions—temperature, water pH and oxygen levels—the fat in the body can, instead of putrifying, saponify and turn into adipocere, also known colloquially as grave wax, forming a permanent cast of fatty tissues. 'Brienzi', a headless male body fully encased in adipocere, was found floating in a bay of the Brienzer See in Switzerland in 1996. Analysis finally determined that he had drowned in the lake in the 1700s and his body had become covered in sediment. Two weak earthquakes in the area might have been sufficient to eventually dislodge him from his incarceration, allowing him to rise to the surface.

Some researchers have called for the building of additional human taphonomic facilities—more colloquially, and more repulsively, known as 'body farms'—where remains are left out in the open air and studied with the aim of providing researchers with a better understanding of the decomposition process. The US has six such facilities and there is one in Australia now, too, but I am not a supporter of the idea of a UK body farm. The arguments put forward to justify them do not sit comfortably with me. There is little evidence that the current method of using animal proxies, generally dead pigs, is inaccurate in establishing a TDI, or that the research from these human-based facilities has significantly improved our ability to predict it with any greater reliability. I would want proof of both to reconsider my position. I find the concept both gruesome and grim, and my unease is heightened when I am invited to take a tour of one of these places as if it were a tourist attraction. I am often asked why we don't have a 'body farm' in the UK. I think the more relevant question is why would we need, or want, one?

◊

Whatever remains of our presence on earth, in death, our identity may be as important as it is in life. Our name—the very core of what we consider to be 'me'—can survive long after even our bones are gone, commemorated perhaps at our final resting place on a headstone, plaque or in a book of remembrance. It may be one of the least permanent constituents of our identity, yet it can outlive our mortal remains by centuries, and in some cases remain sufficiently powerful to inspire fear, loathing, love and loyalty in future generations.

A nameless body is one of the biggest problems for any police investigation into a death, and there is always an imperative to resolve it, regardless of how much time has elapsed between the death and the discovery of the body. Forensic scientists will attempt to link the physical remains with a name so that documentary evidence can be accessed, relatives or friends found to confirm the identification and the circumstances surrounding the death explored. Until this connection is made, there can be no questioning of a person's family, social circle or colleagues, no tracing of mobile phone activity, no examination of CCTV footage, no reconstruction of final journeys. Given the number of people who are reported missing every year—approximately 150,000 in the UK alone—it is no easy task. At its most basic level, our mission is to try to reunite a body with the name it was given at birth.

Generally, we will all have a name—if only our family name—before we are born. If not, we will be given one very soon thereafter. We do not choose it, nor has it been acquired by accident, and we are highly unlikely to be its first or sole owner. This marker, selected by someone else as a gift, or perhaps a curse, for us to carry for the rest of our lives, becomes a significant component of who we believe ourselves to be.

We respond to our names automatically and without hesitation, even at a subconscious level. In a noisy room, we might have difficulty following conversations but when our name is uttered, we hear it loud and clear. Very swiftly, it becomes an embedded aspect of the history of the 'self' we lay down as our life progresses and we may devote some effort, and occasionally significant sums of money, to protecting it from being misused or misappropriated by others.

And yet, despite the importance of our names to our identity, we routinely change them for all sorts of reasons—when we form new partnerships or families, or to separate a personal life from a professional one, or simply because we don't like our birth name. Some people keep one name all their lives; others use two concurrently for different roles or go by a variety of them. Normally, when people choose to formally change their name, the transition is mapped out in traceable official documents; even so, it creates an additional layer of inquiry for forensic investigators.

When you factor in nicknames and abbreviations, one person can be known by a staggering array of labels. My own case is not untypical. I was born Susan Margaret Gunn. As a child I was Susan—or Susan Margaret, the full Sunday name, when I was in trouble, which was quite often. As I grew up my friends called me Sue. I married and became Sue MacLaughlin (Mrs, later Dr); then, on my remarriage, Sue Black (Professor, later Dame)—and for a very short time, to maintain publication continuity for my career, I was Sue MacLaughlin-Black (talk about an identity crisis).

Had my mother had her way, I would have been Penelope, for no other reason than that she liked the name Penny. Not only did I dodge the bullet of becoming known as Penny Gunn, I am grateful that this future forensic scientist did not have to live with the name Iona, lovely though it is when appended to the right surname. Happily, Susan Gunn seemed innocuous enough, although my family name was always going to be a target for exploitation when my initials came into play. Inevitably, perhaps, S.M. Gunn spawned the nickname Sub-Machine Gunn.

Since unique names are rare, most of us will share our most personal label with many others. Of over 700,000 Smiths in the UK, some 4,500 of them are called John. My own maiden name is not particularly common: last time I checked, there were only 16,446 fellow Gunns registered in the UK, most of them, not surprisingly, in the very north-east of Scotland around Wick and Thurso. Only about forty were called Susan.

Encountering a namesake can be amusing but obviously it can cause confusion. For actors, trying to pick a name nobody else is using to secure their Equity card must be a nightmare. When I acquired the name Black another Sue Black emerged on my horizon—a computer scientist who was instrumental in rescuing Bletchley Park from decline. Dr Sue Black OBE is a lovely lady of around my vintage. Although we have never met we have communicated by email, as on occasion I will get queries about Bletchley Park or invitations to give a lecture on codebreaking during the Second World War, and have to inform my very disappointed correspondents that they have 'the wrong Sue Black' and that, unless they want a talk on the dead, they might be best advised to speak to the real one.

Our fascination with identity is reflected the world over in folklore and literary tradition, where stories featuring disguise, assumed identity, mistaken identity or identity theft abound, not to mention foundlings adopted or exchanged at birth. Such themes are a feature of most of Shakespeare's comedies; indeed, much of his work deals with the concept of identity in one way or another. They provide endless plot devices for exploring the nature of society, conflict and how human beings relate to each other.

These stories would have resonated more plausibly in the simpler societies of the past, where creating a new identity, or assuming one belonging to someone else, could be achieved with far less risk of exposure than would be the case today. The infamous sixteenth-century imposter who stole the identity of Martin Guerre, and who has inspired various books, films and musicals, could not have got away with it for so long in the modern era, where forensic science can confirm an identity almost to the exclusion of all others.

Yet there are still plenty of instances where a proverbial skeleton has rattled its way out of a family closet. To find out after many years that you are not who you thought you were can come as a tremendous shock and precipitate a genuine crisis of identity. Was my mother actually my sister? Was my father not my father? Was my father my grandfather? Was I adopted? Since our identity is built throughout our lives on the foundations laid down by those around us—those we trust to tell us the truth—our name and our heritage become the bedrock of our sense of self and our security. It can prove to be a house of cards. When a lie is exposed, everything we believe about ourselves and our place in the world can tumble down around our ears. Such discoveries are often triggered by a death, as relatives search through documents or investigators dig deep into a life to put a name to a body, or to try to understand the circumstances or motivations involved in how it ended.

◊

So when forensic anthropologists are faced with an unidentified body, how do we go about reuniting the deceased with their name? First we need to establish a biological profile: was this person male or female? How old were they when they died? What is their ancestral origin? How tall were they? The answers to these questions allow us to place an individual into a specific pigeonhole. Once we know that we are looking at a woman who was in her mid-twenties, black and around 5ft 6ins tall, we can search databases of missing people to see who might fit those broad criteria. There will be many candidates. One search performed for a white male, aged twenty to thirty and between 5ft 6ins and 5ft 8ins in height, returned 1,500 potential names in the UK alone.

There are three features recognised by the International Criminal Police Organisation (INTERPOL) as primary indicators of identity: DNA, fingerprints and dentition. While fingerprints and forensic dentistry have been used in forensic science for over a hundred years, DNA analysis, the new kid on the block, has been in our forensic toolbox only since the 1980s. We owe its practical application and its revolutionary impact on identification in police investigations, paternity disputes and immigration issues to the pioneering work of the British geneticist Sir Alec Jeffreys from the University of Leicester.

DNA, or deoxyribonucleic acid, is the genetic building block housed within most cells of the body. As half of our DNA is passed to us from our mother and the other half from our father, it has direct familial traceability. There is a common misconception that the recovery of DNA from a body will in itself always lead to a positive identification. But of course a comparison must be made, either with a source sample of the DNA of the person the deceased is suspected to be, if there is one on record, or with samples provided by direct family members (parents, siblings or offspring). The genetic link with a parent's DNA would be equally as strong in, say, an estranged brother of the deceased, so if a family source is to be used it must be supported by other features of biological evidence specific to the missing person, such as their dental records.

When testing parents, we prefer if possible to use a mother's DNA, as obviously there can be some doubt as to whether Dad is the natural father. Families come in all shapes and sizes and biological relationships are not a matter for secrecy in many households, but as there are those where such a revelation could cause great upset, caution and discretion are always at the forefront of such investigations. As my worldly-wise granny used to say, 'You always know who your mother is, but you only have her word for who your father might have been.' That perhaps says quite a lot about my family. Whatever the circumstances, nobody needs the additional burden of unwelcome revelations when trying to come to terms with a bereavement.

A recent mass fatality, in which over fifty people lost their lives, provided a textbook example of how death and DNA analysis can expose family secrets. Two sisters were convinced that their brother had been caught up in the disaster, even though all hospitals had been checked and he was not registered as a patient at any of the accident and emergency facilities. They had not heard from him, and nor had any of his colleagues or friends; he was not responding to his mobile phone and neither were any calls being made from it. Over a week later, there had been no withdrawal from his bank account or use of any of his credit cards.

There was one unidentified, badly fragmented body in the mortuary that fitted this man's general physical description, but the DNA did not match that of his sisters. Investigation would later reveal that the body was indeed their missing brother. Unbeknown to them, and perhaps to him, he had been adopted as a baby—a secret that was eventually confirmed by an elderly aunt. The sisters now had to deal with a double blow: the loss of their brother and the discovery that he was not their biological sibling. For them this raised disquiet about the identity of their sibling, their relationship to him and the honesty of their parents.

UK police forces receive, on average, 300,000 calls relating to missing persons every year—nearly 600 reports a day. About half of this number will go on to be officially recorded as missing, of which around 11 per cent will be classified as high-risk and vulnerable. Over 50 per cent will be between the ages of twelve and seventeen and many of these will fall into the 'absent' or 'run-away' category. A small majority (about 57 per cent) will be girls. Mercifully, many children return or are found alive but more than 16,000 will remain 'lost' for a year or more. When adults go missing, the balance shifts: around 62 per cent are men, most commonly between the ages of twenty-two and thirty-nine, and of the 250 or so people a year found dead in suspicious circumstances, fewer than thirty are children.

The UK Missing Persons Bureau lies within the remit of the National Crime Agency, which has links with INTERPOL, EUROPOL and other international organisations. When someone goes missing, INTERPOL posts what are called 'yellow notices' across its 192 member countries to alert their police forces. 'Black notices' are issued when a body is found and cannot be identified. In an ideal world, all the black notices would correspond to a yellow notice. We attempt to match them by comparing features of identity from the missing person (antemortem) with those of the deceased (postmortem).

The obvious starting points for the collection of antemortem data are the existing national police-controlled DNA and fingerprint databases. However, the deceased will only be represented here if he or she has come to the attention of the police (DNA is also held on different databases for all active forensic investigators, police officers, the armed forces and others, either for identification purposes or to exclude them when samples from a crime scene are being analysed). Via INTERPOL, we can ask permission for other international law enforcement agencies to search their databases if we have reason to believe such a search may prove productive. Most countries do not have universal records of the DNA or fingerprints of the general population and there is no nationwide database for dental records. So unless you are in the police, the military or have been previously convicted of a crime, it is highly unlikely that your identifying features will appear on any database.

Let's take as an example the skeletal remains of the young, white man mentioned on page 52, for whom the missing persons database returned 1,500 possible matches. He had been found in remote woodland in the north of Scotland by a man walking his dog. The police and a forensic anthropologist attended. The bones were lying on the surface of the woodland floor, roughly in the correct anatomical position, although the skull was down at the feet. Hanging from one of the branches of a tall Scots pine tree above the body was the hood of a jacket containing a human bone—the second cervical vertebra from the neck. The body below was missing this vertebra, which proved to be a good fit with the skeleton. It was therefore reasonable to suppose that the body had been hanging from the tree and that, as it decomposed, the tissues of the neck had stretched and eventually given way. The body had fallen to the ground, the head going in a slightly different direction because of the separation of tissues, and the neck bone had fallen into the hood.

All the indications were that the death was not suspicious and was most likely suicide. For whatever reason, it seemed this person had climbed up the tree, tied the hood of the jacket around a branch and jumped off. But we now needed to attempt to identify the deceased so that we could investigate the death properly and inform the closest relatives.

There was no circumstantial evidence of identity. No wallet, no driving licence, no bank cards. We extracted DNA from the bones but there was no match on the DNA database. Since the remains were skeletonised, there were no fingerprints. Our anthropological assessment revealed that the body was that of a white male between twenty and thirty years of age, 5ft 6ins to 5ft 8ins in height.

From his skeleton, we were able to identify some old injuries that had fully healed by the time he died: fractures to three of his ribs on the right-hand side; a fracture of the right collarbone and another to his right kneecap. If all these injuries were sustained in the same incident it was highly likely that he would have been treated at a hospital and there would be medical records. He'd also had four teeth taken out, the first premolar on each side from both the upper and lower jaws. The drift of the remaining teeth showed that these were unlikely to have been congenitally absent but had been removed professionally. Somewhere, then, a dentist would have records of these specific extractions. But we would have to find them.

It was these basic characteristics that generated those 1,500 potential identities. Clearly the police cannot pursue such a vast quantity of vague leads as the drain on resources would be immense. To give them something they can work with, we need to reduce the number of possibilities to double, or preferably single, digits. We undertook a facial reconstruction to recreate the man's features from the contours of his skull. The aim of this process, an ingenious blend of science and art, is not to produce a perfect replica of a deceased person's face but a likeness close enough to be potentially recognisable by those who knew him or may have seen him and thus to generate more precise leads for the police to follow.

The face was reproduced on posters displayed around the area where the body had been found and circulated more widely via newspapers, television, a missing persons website and INTERPOL. After the case was covered by the BBC television programme *Crimewatch*, several strong leads emerged, many of them pointing to the same individual. One of the calls was from his mother. She happened to have been watching the programme and the facial reconstruction reminded her of her son: her worst nightmare imaginable.

With a name to eliminate or confirm, an investigation can

be shifted up a gear from broad physical identity to the realm of possible personal identity. The police can begin to interview relatives and obtain DNA samples for comparison. In this case, Mum's DNA produced a positive match, as did her son's biological profile—white, 5ft 7ins tall and twenty-two years of age when last seen—his dental records, GP's notes, hospital records and radiographs. He had got into a fight several years before he went missing and his broken bones had all been documented in the hospital.

There was indeed no crime to investigate. The man had left home about three years before his body was found, telling his family that he was going to lie low for a while because he had got into some trouble and owed money to a drug supplier. He'd said he was going to head up north and that they shouldn't worry, he'd be fine. He was known in the place where he died as something of a recluse with drink and drug habits, and by a different abbreviation of his name from the one he'd gone by at home.

That a young man chose to end his life is very sad. It is not our place to speculate about or judge what led him to commit suicide, but by giving him back his name, we allowed his story to be told. We were able to provide answers for a distraught family and to return his body to them. The news we bring to relatives is rarely happy, but we believe it is delivered with a kindness, honesty and respect that will ultimately help to set in motion a coping and healing process.

There is no doubt that had our young suicide been carrying some identifying documentation we would have concluded that particular case more swiftly. While most people usually have something on them that gives a clue as to who they are, or at least provides a starting point for an investigation, records such as a universal DNA database or compulsory identity cards would certainly make it easier to identify those who don't. However, the notion of officialdom keeping any closer tabs on us than they do already is controversial and raises concerns for many about the erosion of civil liberties and the right to privacy.

We view our identity as being intimate to us but in reality, we share its finer details with everyone with whom we interact. And every now and again someone acting in an official capacity will want you to share it with them—in our case, when you are no longer alive.

A conversation in *The Death Ship*, written in 1926, between the protagonist and a law enforcement officer sums this up. The author, B. Traven, is an apt contributor to musings on identity as he was himself something of a mystery man. He used a pen name, and his true identity, and indeed more or less every detail of his life, are still hotly disputed.

> 'You ought to have some papers to show who you are,' the police officer advised me.
>
> 'I do not need any paper. I know who I am,' I said.
>
> 'Maybe so. Other people are also interested in knowing who you are.'

CHAPTER 3

Death in the family

'If life must not be taken too seriously, then so neither must death'

Samuel Butlerwriter (1835–1902)

Uncle Willie on Rosemarkie beach.

'Go and check Uncle Willie is OK.'

It was a simple command, casually thrown over his shoulder as my father left the room to attend to the friends and family waiting with my mother and sister in the chapel of the funeral home.

Uncle Willie, my surrogate grandfather, had been dead for over three days. I don't think my father was asking me this because he was squeamish himself. As a typical old-fashioned, no-nonsense ex-military Scotsman of his generation, he would not have been fazed by the sight of Willie's body. And as he did not believe that being a girl meant you should be mollycoddled, he probably took the view that, given my chosen field, I was the best person for the task.

I had dissected a number of cadavers by now, and helped to embalm them, but I was still barely out of my teens and of course learning in the dissecting room was an entirely different matter from coming face to face for the first time with the newly dead body of someone I cared about deeply. It simply didn't occur to my father that I was unprepared to encounter the corpse of my favourite great-uncle in the viewing room of a funeral home. I certainly didn't know what he meant by 'OK'. But he had given me the job and we always did what Father told us—it would never have crossed my mind to say that I didn't want to. My father always barked his orders as if national service had never ended and his sergeant-major's moustache still bristled with an authority that brooked no dissent.

Willie had been a huge presence in more ways than one. A jovial man of wide girth, he had not a single grey hair on his head when he died at the respectably good age of eighty-three.

He had fought in the Second World War but, like so many men of his generation, he never talked about it. By trade he was a master plasterer and was responsible for the beautiful and ornate cornicing in many of the grand houses in the more affluent parts of Inverness.

It was a source of great sadness to Willie and his wife Christina, always known as Teenie, that they had no children of their own. So when my maternal grandmother, Teenie's sister, died seven days after giving birth to my mother, they gladly took in the baby, bringing her up in a house full of love and laughter. They were true grandparents to me in every sense: kind, caring and generous to a fault.

In his retirement years Willie used to wash cars at the local garage to make a few extra pennies. I remember him standing in the wash bay, hose in hand, 'Willie's wellies', as he used to call them, on his feet, both turned down on his calves because his legs were too fat for them, cigarette hanging out of the corner of his mouth, always laughing. For some reason he loved blowing raspberries, which made him irresistibly naughty to us children. With help from the family, he cared for his disabled wife throughout the often extreme ravages of her dementia, crippling arthritis and debilitating osteoporosis. He considered it his duty to her, as was the way of many families back then, and would not discuss her going into a hospital or nursing home.

After Teenie died, Willie would come to lunch at our house every Sunday and usually joined us on family outings when the weather was kind. I never saw him outside his own home in anything but a three-piece suit, shirt and tie. He only had two suits, one in heavy tweed for everyday wear, and a best suit for funerals.

There is a photograph of Uncle Willie that epitomises his zest for life and the laughter he spread. It was taken on Rosemarkie beach on the Black Isle, just north of Inverness. It was a baking hot day and we had driven there for a picnic on the beach, all crammed into Father's car, which at that time would have been his two-tone black-and-tan 3.8 Mark II Jaguar, his absolute pride and joy.

Even to eat his sandwiches on the shores of the Moray Firth, Uncle Willie was dressed as if for church, in his suit and perfectly polished shoes. We unfolded one of those lightweight, tubular metal garden chairs on the dry sand and suggested he took a rest in the shade while we set out the blankets and picnic further down the beach. As we busied ourselves with Mother's spread—as usual, enough to feed an entire battalion—there was an explosion of hilarity behind us. Uncle Willie had wedged himself inextricably into the flimsy chair and as his not inconsiderable heft bore down on the spindly frame, it started to buckle and sink into the sand. Like the captain of a ship disappearing under the waves, he raised his hand to his forehead in salute as he descended, rather gracefully, his legs straight out in front of him, until he came to rest on his bottom. The picture shows him roaring with laughter at his farcical plight and you cannot help but smile with him. He had little in life but he was a hugely contented man.

Uncle Willie died in a way that would have made him laugh just as heartily, had he been capable of it. One Sunday, at our house for lunch, he just slumped at the table, as if suddenly dropping off to sleep. He had suffered a ruptured aortic aneurysm, something that strikes without warning—a mercifully instantaneous death for him but a brutal shock for my rather emotional and sensitive mother. One moment he was his usual jolly self and the next he was gone. Unfortunately for Uncle Willie, and for my mother's tablecloth, he collapsed rather inelegantly, face first, into his bowl of Heinz tomato soup. It was as if our uncle was determined to keep his sense of humour to the bitter end.

And now here we were, relatives and friends united in grief

at the funeral home, ready to mourn the passing of the last of a generation. But first I had to take a deep breath, pull up my big-girl pants, do what my father had asked of me and perform the last service I could for Uncle Willie: check he was 'OK'.

I imagine that for everybody, viewing a dead loved one is a moment to pause and take stock of what they were in life, to hold on to that memory and not allow it to be clouded by what they have become in death. Willie had been a kind, gentle soul and an irresistible life force. I never heard him utter a bad word about anybody or complain about anything. This was the man who would let me place pretend bets on horse races, who took me to the shops to buy sweets, who let me help him wash cars—a man who was simply a joy to have in my young life. My only regret was not having the chance to get to know him better as an adult.

I remember the subdued lighting of the viewing room, the vaguely hymnal music playing low over the speakers, the smell of flowers and perhaps a slight whiff of disinfectant. The wooden coffin was raised on the catafalque in the centre, surrounded by flowers, the lid yawning open, waiting to be screwed down for ever so that he could rest in peace.

Registering cataclysmic shock, I realised suddenly, and all too clearly, the enormity of what had been asked of me by my father. The man in the coffin would not be buried until I gave him the all-clear. Uncle Willie had to pass muster. I felt I had been entrusted with an important mission but I was more than a little apprehensive. There was no way of knowing just how prepared I was for this and how it might affect me.

I approached the coffin, hearing my own heartbeat in my ears, and peered inside. But this wasn't Uncle Willie. I took a sharp intake of breath. Lying within the white lining was a much smaller man, the ruddy complexion replaced by a waxy pallor and maybe just a hint of foundation. There were no laughter lines around his eyes, his lips had a blue tinge and he was, inconceivably, silent. For sure, he was wearing Willie's best funeral suit, but the essence of the man had gone and all that was left was a faint physical trace of him in a shell once occupied by a huge personality. I realised that day that when the animation of the person we were is stripped out of the vessel we have used to pilot our way through life, it leaves little more than an echo or a shadow in the physical world.

Of course, it was Uncle Willie in the coffin—or at least, what remained of him. He was just not as I remembered him. It was an experience that would replay in my mind in later years when I witnessed families walking up and down rows of dead bodies laid out on the ground after a mass fatality, searching for the face of someone they desperately wanted, or in many cases did not want, to see there. I recall some of my colleagues being incredulous that people could not recognise the bodies of their closest relatives. But through my own personal encounters with death I had come to understand that the dead, even those you know, look very different from the living. The changes brought about in the appearance of a human body are more profound than can be accounted for simply by the cessation of blood flow and loss of pressure, the relaxation of muscles and the powering down of the brain. Something quite inexplicable is lost—whether we choose to call it a soul, a personality, humanity or just a presence.

The dead are not as they are depicted in the movies by actors lying perfectly still as if in a deep sleep. There is a void in them which serves somehow to weaken the certainty of the bonds of recognition. Of course the explanation for that is simple—we have never seen them dead before. Dead really is dead, it is not just sleeping or lying motionless.

Back then I couldn't understand my inability to recognise Uncle Willie, and it unsettled me. It wasn't as if I could attribute his appearance to disruption caused by a violent death or decomposition. It hadn't been a violent death, and it had occurred only three days earlier over Mother's soup—Scotland does not hang around to bury her dead.

I reasoned that in a small place like Inverness, where Willie was known to everyone, as were my parents, it was very unlikely that this was ever going to be a question of mistaken identity, still less of someone switching the body in the coffin or doing something illegal to the corpse. He had been born here, grown up here, married here and had now died here. The funeral director was a relative of Willie's, for goodness' sake—he wasn't going to get it wrong. Of course this was Willie. But even though the rational part of my brain knew that, the disconnection between how he'd looked in life and how he looked in death was very perplexing.

After my initial hesitation had passed, I became aware that there was a sense of peace in the room. The silence around the dead has a different quality from the silence that is just an absence or cessation of noise. There was a calmness to it, and the fear that I was going to find myself afraid began to dissipate. Once I realised that the Uncle Willie I had known had truly gone I was comfortable with what was left of him, though I understood that my relationship with him now needed to be different from the one I had with the cadavers in my dissecting room. Those I knew only on one level, in their present state as dead bodies, whereas Willie existed for me on two planes: in the present as the physical form in front of me in the coffin and in my memory as a living person. The two manifestations of him did not match and there was no reason for them to do so as they were not the same. The man I remembered was Willie. The other was just his dead body.

My duty should have entailed nothing more than a quick glance into the coffin to verify that the man lying there was indeed my great-uncle and that he was suitably dressed and looking smart, as he would have wanted, before being laid to rest. However, in my youthful enthusiasm to do things properly,

I went overboard. I slipped into a pompous analytical mode worthy of *Monty Python's Flying Circus*. No dead parrot in this sketch, though—only poor old dead Uncle Willie.

Had any of the funeral staff walked into the room they would have questioned my sanity and possibly even have had me escorted from the building for disturbing the peace of the dead. Certainly no other corpse in the history of that highly respected Highland funeral home can ever have left the premises with such a rigorous MOT.

First I made sure that he was dead. Yes, really. I felt for a radial pulse at his wrist and the carotid one at his neck. Then I placed the back of my hand on his forehead to check his temperature. How on earth I could imagine he might show any signs of life or warmth after being in the funeral home fridge for three days, I don't know. I noted that there was no bloating of his face, no skin discolouration and no advanced odour of decomposition. I examined the colour of his fingers to ensure that the light embalming fluid had fully taken, and his toes, too (OK, I admit it—I took off one of his shoes). I gently prised open an eyelid at the corner to check that his corneas had not been removed illegally and opened a button of his shirt to rule out any evidence of an improper postmortem incision. I knew one should never overlook the possibility of the theft of body parts. Honestly? In Inverness? Not exactly the epicentre of the international black market in stolen organs. Then, perhaps worst of all, I checked his mouth to establish that his false teeth were in place. Who would have wanted to steal Willie's wallies? One careful owner, free to a good home . . .

Noticing that his watch had stopped, I instinctively wound it up and placed his hands across his large tummy. Did I seriously think he was going to want to know what time it was when he was in the ground at Tomnahurich Cemetery, and perhaps ponder on how long he had been lying there waiting? For what? In the unlikely event that he woke up, he wouldn't have been able to see his watch without a torch anyway and I hadn't thought to bring one of those, had I? I moved an errant lock of Brylcreemed hair that had strayed across his face and patted him gently on the shoulder. I thanked him silently for being who he had been and, with a crystal-clear conscience, returned to my father and reported that all was well with Uncle Willie. He was certified fit to bury.

I crossed many boundaries that day, and without much logical justification. Although I look back on my actions with incredulity, of course I understand now that death and grief do strange things to a mind. It had been a first experience for me and I had handled it in the only way I felt I could. And it was an important milestone. It confirmed that I could compartmentalise: as well as bringing compassion to my dealings with the bodies of strangers, I could manage the emotions and memories involved in viewing the mortal remains of a person I had known and loved while accessing the detachment required to inspect him professionally and impartially without falling apart.

In no way did this diminish my grief but it showed me that such a compartmentalisation of emotions was not only possible but permissible. For that lesson, I have Uncle Willie to thank and also my father, who simply assumed that this was a service I was equipped to perform and did not doubt for a moment my ability to do it. And I am glad that I did.

My reward from Father was a curt nod of the head that told me he accepted my word. From that moment on, I have experienced no fear of death.

◊

Fear of death is often a justifiable fear of the unknown; of circumstances beyond our personal control which we cannot know and for which we cannot prepare. '*Pompa mortis magis terret, quam mors ipsa,*' the philosopher Francis Bacon wrote over 400 years ago, quoting the Roman Stoic Seneca. 'It is the accompaniments of death that are frightful rather than death itself.' Yet the control that we like to think we have over our lives is often an illusion. Our greatest conflicts and barriers exist in our minds and in the way we deal with our fears. It is pointless even to try to control that which cannot be controlled. What we can manage is how we approach and respond to our uncertainties.

To understand the roots of the fear of death, we may need to unpack it into three stages: dying, death and being dead. Being dead is probably the least bothersome as most of us accept that this is not something from which you recover and that worrying about the inevitable is somewhat futile.

Such fears we may have of being dead will depend on our concept of what happens to us afterwards: believers in versions of heaven, hell or some form of survival of the soul may have a different perspective from those who anticipate oblivion. Death is a genuinely unexplored destination for which, as far as we know, there is no return ticket. Certainly nobody has ever come back with credible scientific and verifiable evidence that they have actually been there. Of course, very occasionally someone believed to be dead does start breathing again, but given that over 153,000 people die every day on this planet, I suspect that the sample size of those who have 'come back' doesn't reach statistical significance, and no further real scientific understanding has been gained from such cases.

We have all heard stories of near-death experiences, portrayed as mystical events involving floating, out-of-body sensations, bright lights and tunnels, visions of previous episodes in the person's life and a feeling of calm. They tease us with the possibility that we can know what death will be like; perhaps even that we can defy it. Science has alternative explanations. All of the phenomena reported can occur normally if the right biochemical conditions or neurological stimuli exist to impact on brain activity. Stimulation of the temporo-parietal junction on the right side of the brain will generate a sense of floating and out-of-body levitation. Vivid imagery, false memories and the replaying of real scenes from the past can be induced by fluctuating levels of the neurotransmitter dopamine, which interacts with the hypothalamus, amygdala and hippocampus. Depletion of oxygen and increased levels of carbon dioxide can cause the visual hallucination of bright light and tunnel vision, as well as a feeling of euphoria and peace.

Stimulation of the fronto-temporo-parietal circuitry of the brain can convince us that we are dead—even that we are drained of blood, devoid of internal organs and decomposing, in the case of sufferers of the rare psychiatric disorder Cotard's syndrome.

It is human nature to prefer a mystical or supernatural explanation rather than trust the logic of biology and chemistry. Indeed, this is the premise on which all bogus mystics and prognosticators rely when promoting their smoke-and-mirrors act to a vulnerable client.

The greatest fear tends to be focused on the manner of death—the dying bit. The precarious and painful period, be it moments or months, between the point when we know death is upon us and when it actually occurs. Will we live out our final days with illness, be snuffed out suddenly through accident or an act of violence, or simply fade away? In short, will we suffer? As the writer and scientist Isaac Asimov put it: 'Life is pleasant, death is peaceful. It's the transition that's troublesome.'

Would that we could all be as lucky as Uncle Willie and meet the end of a long, happy, healthy life in a sudden, pain-free collision with a bowl of warm tomato soup, surrounded by our family. He had no fear of dying as he did not know it was coming. For me, this is the perfect death, the kind I would wish for anyone I loved. In the short term it is a shock for the bereaved. My mother had no time to prepare for the sudden removal of the man who was effectively her father, no time to gear up for her own grieving process. The dying ritual she had expected had been denied her and the being dead part had come without warning. In the long term, though, those left behind are invariably comforted to know that the person doing the dying did so in the least physically and psychologically distressing way possible.

A jovial uncle who loved his food collapsing during lunch; a gardener struck by a heart attack landing face-first in manure . . . death and black humour are old companions. Even if the capriciousness and frivolity of death are rarely funny at the time, they can provide those left behind with a much-needed coping mechanism. Cold irony can be more cruel: the proud, independent man who always feared incapacity spending his last years locked inside his body in an impersonal care home; the liver pathologist killed by hepatic cancer; the solitary death in a hospital bed of a woman scared of dying alone . . . These are all fates that have befallen some of my own friends and relatives.

My beloved grandmother, who was a *teuchter*—a Gaelic-speaking Highlander—believed completely in the second sight. She told many tales about her own grandmother who, she said, could predict when somebody from their small west-coast community was walking towards their *caochladh* (end of life) because she would dream about their funeral. Great-Great-Grandmother knew whose death she was foreseeing because she would recognise the chief mourner in her dream.

One of these stories concerned 'Katie up the Glen', a distant relative of my grandmother's, whose approaching demise the old lady foretold after seeing Katie's funeral cortege, led by her husband Alec, in one of these dreams. This was a bit of a shock to all and sundry as Katie was not old and extremely hale and hearty. But as spring turned to summer my great-great-grandmother was adamant, even warning that Katie's end would not be long in coming: in her dream she had seen the peats being cut, which meant summer was on its way. Every day poor Katie was monitored closely and every day she went about her business without complaint or ailment. When the peat-cutting began Katie was out there with everyone else, hauling the slabs up on to the bank and leaving them there to dry before they were transported back to the croft by a cow and cart. The black swarms of midges would have been merciless and the work back-breaking.

Whatever startled the heilan coo that day nobody could say, but poor 'Katie up the Glen' got caught between the beast and a dry-stone wall and was crushed to death. As predicted, Alec did indeed walk behind her coffin to the graveyard that summer. My grandmother was always mischievous and I wouldn't put it past her to have made up the whole story. If she didn't, it seems highly probable that some of the women in our family would have been burned as witches in the past—especially the ones with red hair. Superstitions like this form part of the untameable root system of the many misconceptions surrounding death. But they do make great spine-chilling stories for scaring the children on cold winter's nights around a peat fire.

My grandmother—my father's mother—being of a generation who often died at a much younger age than we do today, was the only one of my grandparents I ever knew and the most important person in my life. She was my teacher, friend and confidante. She believed in me and understood me when nobody else did, and whenever I needed the advice, conversation or reassurance of a grown-up one step removed from my parents, she was there for me. Even when I was a child, she talked to me honestly about life, death and being dead. She had absolutely no fear of it. I often wonder whether she saw her own death coming. I recall, during one of our more memorable dark talks, experiencing a moment of clarity when it struck me that she would not always be there with me, and it made me very sad and very scared. I didn't want ever to lose her.

My grandmother looked at me sternly with her deep black eyes and told me that I was being *faoin* (silly). She was never going to leave me, even when she went 'beyond', as she used to call it. She vowed that she would always be sitting on my left shoulder and that if I ever needed her for anything, all I would have to do was turn my ear to her and listen. I never doubted her, and I never forgot her promise. Indeed, I have lived with it, and by it, every day of my life. I still automatically tilt my head to the left when I am thinking, and I can still hear her voice giving me advice when I need it. I am not sure even now whether it was a kindness to a frightened little girl or a curse, because I could have had so much more fun growing up if it hadn't been for my dead grandmother. There are many times when she has stopped me from doing something I've wanted to do but knew I shouldn't. Some might call it conscience, but there is no doubt that my own little Jiminy Cricket speaks in my grandmother's lilting, Highland voice.

In that same conversation, she made me promise that I would look after my father, her only child, when his time came. Nobody, she said, should have to step through death's door on their own. She would be there waiting for him on the other side but I must be the one to walk him to the threshold. I never questioned such an odd request—I was only ten. I never asked, either, why my mother would not be there for him. And as it turned out, she wasn't. Did my grandmother, long gone by then, somehow foresee that my father would be the last of his generation to die, with only the next generation there to look after him on his way?

Dying is a path we don't have to walk unaccompanied, but when we reach that door we cross the threshold alone. Our myths, fables and culture instil in us notions of how death will be and what we should expect, but where is the evidence of how it will be for me, or for you? It is an incredibly intimate and personal transition—the end of everything we know, are and understand, and no textbook or documentary can prepare us. If we cannot influence it, perhaps we shouldn't waste precious time worrying about it. When it comes, we just need to experience it.

My grandmother died in an impersonal hospital bed. A heavy smoker, she'd had investigative surgery for chest pains and when they opened her up they found lung cancer so rife that there was nothing to be done and just closed her up again, very quickly. I knew it was not the death she would have wanted but at that time, with such an illness, there was little alternative to a clinical, medicalised death in hospital. There was no opportunity for her to be at home, no comfort and quiet. As children, we were not encouraged to visit her in hospital and so I never saw her again. It is a deep regret that has stayed with me all my life. I wish I had been able to talk to her even one last time, to hear what she had to tell me about her dying and death and to learn from her wisdom.

And so my first real experience of death, at the age of fifteen, was the loss of the person who meant most to me in the world. My father, mindful of how special our relationship had been, asked me if I wanted to see my grandmother in her coffin. Hurting badly and scared to look at her lifeless body, I declined—to the relief of my mother, who had not been happy about it in the first place. I bitterly regret that, too. It is a source of great sadness that I did not have my last moment with my grandmother, just the two of us, either when she was dying or following her death. Perhaps that accounts for the overcompensation with Uncle Willie.

What we could do was celebrate her life—and boy, did we. My mother cooked until there was nothing left in the cupboards, the whisky and sherry flowed and the windows of our sitting room were flung open to let her soul fly. The last image

I remember from that day is the sight of our minister doing an eightsome reel in the front garden with the music blaring out from our stereo. Yes, it was some party we had, and she would have loved it. I wonder if she felt she was going to meet her maker. We were not an overly religious family, although we did attend church and held strongly to Christian values. I remember my grandmother having some fierce philosophical debates with the local minister when they were playing cards. While he was deep in thought, she would be switching the cards right under his nose.

She was such a staunch believer in a life after death that I almost wished she would come back and tell me what it was like. Sadly, she never has.

CHAPTER 4

Death up close and personal

'Sometimes you will never know the value of a moment until it becomes a memory'

Theodor Seuss Geiselwriter,
cartoonist and animator (1904–91)

My mother and father, Isabel and Alasdair Gunn, on their wedding day in 1955.

Almost all of us will encounter death at close quarters long before facing her ourselves, and those experiences can have a profound influence on our fears, attitudes and on what we perceive as a 'good' way to die. For many of us, the first of those most intimate vicarious encounters to really hit home is likely to come with the death of our parents.

As adults, we accept that it is our responsibility to manage the dying and death of those who brought us into the world, as it has been for the second generation since time immemorial. It is in the natural order of things that children bury their parents and not the other way around. Today, with people living longer, and sometimes in more complex families, it is of course more commonplace for multiple generations to be living simultaneously. When 'children' might be in their seventies, the responsibility for managing the deaths of both grandparents and parents may fall to the third generation. Whatever the circumstances, as well as obliging us to get to grips with the realities of death, and often to introduce our own children to her, the loss of our parents is a reminder that we are ourselves growing older and that can bring with it a sharp focus on our own mortality.

Since most of us are products of a culture and era where nobody likes to discuss death in case it encourages her to visit, it can be difficult to know what our loved ones want to happen when their time comes, and indeed how we should go about getting ready for that. My husband, Tom, and I often used to talk about which of our four parents we thought would go first and who would outlive the rest, joking that the creaking gate always hangs the longest. But this was no morbid parlour game:

it was an attempt to plan for the management of elderly lives in such a way as to maintain dignity and independence for as long as possible. In the event our great predictions were totally wrong. The one we thought would go first, my father, outlasted all the others by quite a few years—and even he would have admitted that only the good die young.

Did I fear my parents' deaths? In truth, I don't know. I think that, beyond being concerned about what dying might entail for them, I was not preoccupied by the prospect of their actual deaths or of them being dead. I saw their demise as an inevitability for which pragmatic planning was essential. I don't intend to sound cold—I adored them both and would dearly have loved them to have lived the longest possible healthy and happy lives—but as death is a certainty, we need to be prepared for it.

My mother took ill rather suddenly. I was teaching police in a week-long training programme when a call came through from my father to alert me that she had been taken into hospital. As I'd expected, he was totally unhelpful in terms of being able to give me any real information. I finished my part of the programme and drove from Dundee to Inverness. The A9, bogged down by lorries, caravans and tourists, can be a long, lonely and frustrating road when you need to be at the other end of it in a hurry without risking licence and life.

When I arrived on the ward my mother's first words to me were 'You came.' She had always been afraid that if her health failed nobody would want to take care of her and that she would be left alone. Having spent her entire life looking after other people—her aunts and uncles as she grew up, her husband and then her own family—she placed such little faith in her own value that she was unable to accept how much of a role model she was to us all. Now it was my job to look after her. She had suffered from hepatitis as a young woman and her liver was now slowly shutting down. Other organs were also failing; ascites, an accumulation of fluid in the peritoneal cavity, was becoming a problem and increased levels of bilirubin were producing jaundice. At her age, my mother was not going to recover.

She had never managed to make the transition comfortably from the mother-child relationship to its mother-adult daughter phase and we rarely held deep grown-up conversations. As a result, she really knew little about me, found me impenetrable at times and was therefore reluctant to share either her fears or her hopes. Ours was not a talkative, open and sharing family in general and my mother would have found it embarrassing to discuss her personal needs with anyone. Teenie and Willie had done a tremendous job with the little girl who had lost both her parents but, having been so sheltered and cosseted, she grew into a very dependent woman. I, on the other hand, had inherited my father's and grandmother's no-nonsense, independent approach to life and I was conscious that my mother found me hard to get close to and to understand. But she also knew that when things were bad, she could always turn to me as I would be logical and practical and I would cope.

Now, faced with her rapidly deteriorating health, I felt that she would not wish me to probe into what she might or might not want me to do for her. For her part, she expressed no desire to endure any medical interventions that might have postponed her decline and did not ask me to help her prolong her life. It seemed my mother had accepted that her time had come and had found a personal accommodation with it that appeared not to dwell on either regret or unrealistic expectation. My instinct was that, as she had so often done in the past, she was putting me in control of decision-making, this time for what would be the remainder of her life. My father and my sister were relieved, as neither wished to take on the responsibility. I undertook to do what I could to manage her dying and, ultimately, her death and the necessary rituals. I did it willingly, though with a heavy heart, and proudly, as the last service a grateful daughter could perform in this life for a genuinely kind and loving mother.

I remember the relief on the house doctor's face when I stated very firmly to him that I did not want my mother to be resuscitated should the need arise, or to be placed on a drip. Nor did I want her listed on a transplant register. All these were notional lifelines the registrar was duty-bound to offer families as a last vestige of hope, even though, in reality, as we both knew, there was no realistic hope. The only effect any of them could have had would have been to extend my mother's dying. To use an organ that could have profoundly benefited a younger person would have been unconscionable for both of us. This was something of which I was certain because my mother had in the past voiced the view that, with organs in such short supply, transplanting them into old people was a waste.

I did manage to get her home to her own bed for just one night before she died, but it took a great deal out of her and distressed her terribly. She was horrified that she had been catheterised and needed help to cope. I remember asking her whether, if the tables were turned and it was me who needed assistance, she would do this for me. The question was brushed away with irritation—of course she would. She was forced to concede, however unwillingly, that sometimes the roles of parent and child need to be reversed. When I got my mother back to the hospital the next day, it was clear she was not ever going to be able to go home again. She needed a level of palliative care that only the hospital could give her, or so the culture of our health system led me to believe. In any event, I allowed her death to be medicalised, and left the doctors and nurses to perform the intimate tasks she would have hated being done by anyone, let alone strangers.

I may have been broadly in charge of the decisions and directives passed on to the medical team, but it was they who would dictate the pace of her dying and control the level of her engagement with the world around her. In my moments of dark reflection, I chastise myself for those hours she spent alone in the hospital. Friends would visit early on but gradually they fell away as she became less responsive. I believe she would rather have been at home, where she would have been loved and cared for in her final days, but my father could never have coped and there was not the same level of home nursing available then as there is today.

In our busy lives we try to juggle what we think we should be doing with what we must do and what we want to do. In the end, most of us will probably feel that we didn't achieve enough or should have done things differently. Yes, I had a husband, children and a demanding job over a hundred miles away, but I had only one mother—a mother who had always lacked self-esteem and, although kind of spirit and heart, was fundamentally sad, lonely and unfulfilled. So I regret simply accepting it as the 'norm' that she would be cared for in a hospital and that, in my absence, she might be visited by others. Would I do it differently today? Yes, I probably would, but that perspective comes with hindsight and experience. As the older generation of my family have died, one by one, I think I have got better at managing the process with them. Practice makes perfect, or so they say.

There were only five weekends between my mother's first admission to hospital and her death and my daughters and I spent every one of those with her in a tight little family bubble, cramming in as much time as possible with her while we could. On our penultimate visit she was slipping into a coma. I told her we would be back again the following Saturday and that she should hold on until then, although I had little faith that she would. Such arrogance to expect her to arrange her dying timetable around us! At the time, I felt it was the right thing to say, to encourage her to have something to look forward to (utterly insane: the woman was dying), but I wonder if I merely extended her suffering and loneliness. I shiver now at my thoughtlessness. I am ashamed that I allowed my dogmatic personality to assume control in the expectation that she would simply comply; that I took it for granted there was some benefit to her when, in reality, there was none. Maybe I am being hard on myself, but nobody will ever convince me that she did not hang on for us to visit her one last time when she could have been at peace sooner.

A hospital ward, devoid of warmth, love, character and memories, can be such a sterile environment for the dying and their loved ones to try to prepare themselves for the most personal, private and irreversible of moments. The next Saturday, the last time I saw my mother alive, my two youngest daughters and I spent the afternoon alone with her, largely uninterrupted. I was certain this would be their final chance to say goodbye and I didn't want them to grow up regretting, as I had always done, the loss of those precious last hours with their grandmother.

My mother was in a side room on her own, now in a morphine-induced coma and no longer with us. Or was she? The auxiliary nurse taking care of her final needs was simply going through the motions. She wasn't in any way cruel or neglectful, but she showed no empathy or understanding for either my mother or for us. She had a job to do and we were almost irrelevant.

Our middle daughter Grace, who was twelve at the time, was incandescent with rage over such a lack of compassion. Her anger and indignation never left her—indeed, it was instrumental in our clever little monkey of a daughter becoming a nurse herself. Experiences of death have the power to alter attitudes and even to change the course of lives. Grace has buckets of understanding and a huge heart, qualities that make her exactly the kind of nurse her grandmother should have had in her last hours on this earth—the kind of nurse every family is entitled to expect. She is not afraid to sit and hold a patient's hand in their final moments, offering comfort and reassurance untainted by falsehoods. Isn't that what we would all want when we are ill, in pain or dying—kindness and honesty? It does not surprise me that recently she has been considering specialising in palliative care. It will be a heartbreaking road, if she chooses to take it, but I know she would fight for the dignity of every single patient in her care. Her grandmother would be as proud of her as we are. Yes, Grace is our own angel of mercy—even if she does currently have blue hair, which must terrify some of her poor patients.

Research using electroencephalograms (EEG) suggests that of all our senses, hearing is the last to go when we are unconscious or dying. This is why palliative care professionals are very cautious about what is said in the vicinity of a patient and why families are encouraged to talk to those who appear comatose. That last weekend, we decided that Granny should not leave this world hearing nothing but silence punctuated by distant whispers and tears. We would not be a family that moped, we would be the Von Trapp family: we would sing.

Even though recalling her death is still sad and painful, the memory of that bizarre last day still makes the girls laugh. We went through our repertoire of hits from Disney films, a range of Christmas songs (despite it being the height of summer), all my mother's favourites and one or two old-fashioned Scottish ditties. Every time a nurse or doctor came into the room they smiled and shook their heads at the sight of the three of us lolling around belting out songs in inglorious disharmony. The look on their faces would catapult us into further hysterical merriment and the room filled with love, laughter, light and warmth as well as caterwauling. Hospitals are terribly unhealthy places for the soul and bringing more laughter into them can only be a good thing. There were no ministrations from clergymen, no grief-stricken friends—just 'her girls' having a fun time, keeping her company and simply being human.

Death is, after all, a normal part of life and sometimes in Western cultures we hide it away when maybe what we need to do is to embrace it and celebrate it. Sometimes, with the best of intentions, we try to shield our children from harsh realities when perhaps we could be preparing them to face the events they will need to deal with in the future. I know that not everybody agrees with that philosophy, but it was important to me that my children were there, not only to say a proper goodbye to their grandmother but also so that, when it is their turn to attend to me and their father, they will know that it is OK to laugh and be silly and that we would rather have laughter and song than heartbreak and tears. Maybe some people would consider it disrespectful to belt out 'The White Cliffs of Dover' and 'Ye Cannae Shove Yer Granny Aff a Bus' round your mother's deathbed, but I think she would have enjoyed it immensely.

After all the classics had been given an airing, we were exhausted. My mother had not moved an inch in the entire time we'd been there but we had held her hand, moistened her lips and combed her hair. It was when the time came to say our farewells that the tears inevitably came to the fore. When the girls had said their goodbyes, I asked them to give me just a moment on my own with my mother. But I found myself unable to utter a word. I couldn't tell her that I loved her, that I would miss her, or even summon the words to thank her. Neither my mother nor my father ever told me they loved me, although I always knew they did. The expression of such sentiments had never been part of our family language, and to voice them now seemed just too alien to the strange, stiff-upper-lip attitude that set the tone for the way our family dealt with each other. Besides, I was afraid that if I did, I might start crying and never stop, and I did not want my daughters ever to see me distressed. My role was to be the strong one.

And so I simply said goodbye, and closed the door on her room, leaving her to make her final journey alone. I regret that decision more than anything. If I could, I would go back and change every single aspect of my farewell. While I will always feel that I should have been with her at the end, I fear that if we had stayed she would have continued to cling on to life for us. I had to let her go, and it seemed to me that it was only by departing that I could do so.

I had just walked into our house, only two hours later, when the hospital phoned with the news that my mother had died. How quickly had that happened? Had she just been waiting until we had gone to slip away? Or had she lain there alone for a while in the silence we left behind us? Perhaps she was glad there was silence finally and we had stopped all that dreadful singing. Somehow I doubt that. Had she been on her own in that hospital room or had a compassionate nurse sat with her in those last moments? Had the morphine just allowed her to pass away quietly and unconsciously?

I will never know the answers to those questions. All I can be sure of is that, though she wasn't able to die in her own bed in her own home with her family around her, as she would have wished, we tried to do the best we could for her. I sincerely hope she understood that. Whatever plans or promises we make, illness and death have a habit of shifting the goalposts.

It can be harder than we think to be with someone when they die. You can maintain a round-the-clock vigil at the bedside of a dying loved one only for them to breathe their last when you are grabbing a couple of hours' rest or have just popped out for a cup of coffee. Death works to her own timetable, not ours.

Tom and I allowed our girls to decide whether they wanted to see their grandmother one last time before her funeral. We didn't want them to go through life afraid of the unknown or feeling that they had been denied this opportunity to help them accept her death. The three of them went into a judge-like huddle and resolved that they would all like to see her. Beth was a grown woman of twenty-three but Grace and Anna were only twelve and ten. The room at the funeral home was quiet and the coffin was open—memories of Uncle Willie came flooding back, but I am glad to report that I behaved this time.

I learned that day to have faith in the resilience, dignity and decorum of our daughters. As I stepped back to permit them their first personal encounter with death, they all commented on how terribly small their grandmother looked. It was, true to form, Anna who made the first move. The fearless little one who used to give her granny near heart failure by clambering to the top of the highest climbing frame in the wildlife park and waving enthusiastically at the people far below on the ground, holding on with just one hand.

Anna leaned into the coffin, took hold of my mother's hand and stroked it gently. No more was needed and no more was said. The touch of love that shows no fear of death. Granny had finished with dying, had died and was now dead—all very clear and distinct concepts in their minds. They were at ease with the finality. They know that the best memorial is a boxful of happy memories inside your head, and they know what a good death looks like.

◊

My father remained strangely detached in the aftermath of my mother's death. He never offered to arrange anything, or took any responsibility—he just seemed almost passively to allow everything to happen around him. He and my mother had been married for fifty years and yet there appeared to be little grief in him. At the time I put this down to a combination of his innate stoicism and shock.

With hindsight, I believe the dementia that would soon

emerge and consume his own life had already begun to take hold, and that my mother had been covering up the changes in him, making all the usual excuses for his forgetfulness and odd behaviour. At her funeral, a traditional, solemn affair, I think he just went through the motions and I am not sure now that he really understood what was going on. The telltale signs were there, but amid the distractions of the bureaucracy of death and our own grieving we just didn't see them, or perhaps we chose not to. He offered no reminiscences, shed not a tear and everything for him seemed to be business as usual. After the service he chatted away to friends and family as if it were a wedding rather than the funeral of the woman with whom he'd shared his life.

If my mother's dying days were mercifully brief, my father's would be painfully protracted and, given a choice, he would certainly have opted for a different final journey. In fact, if he'd known what was coming his way, I have little doubt that this no-nonsense Scot with no time for sentimentality would have gone out into the woods behind the family house with his shotgun and ended it all. I remember often thinking that the kindest way for him to go would be to fall off the roof when he was up there trying to fix a slate. But was it difficult for him or just difficult for the rest of us? How much greater is the burden of grief for those who have to watch Alzheimer's rob someone of their memories, and most of their very identity, than it is for those who have them stolen?

Left to his own devices, what we thought of as abnormal behaviour, exposed by the absence of my mother's interventions, became his reality. He cursed the boys, figments of his imagination, who had come into the house and stolen his keys; he told our girls to keep quiet in case they woke their grandmother, who had been dead by this time for over a year. All of these signs creep up on you, and initially you make allowances and adjustments for them.

We rarely plan for dementia, we just manage it. Tom and I met each new problem with new solutions. My parents had lived in the family home since 1955 and although we had tried on many occasions to get them to downsize, my father would always say, with a twinkle in his eye, that it would be our job to clear the house when they'd gone. We could not move him now. We organised a carer to come in three times a day to make sure he ate properly—we had food delivered that just needed to be heated up—and that he was safe and warm. We made the 230-mile round trip to Inverness almost every weekend to clean and maintain the house, change his bed, do the laundry and shop.

It takes a crisis to force you to face reality and make some major and unpalatable decisions. This came one bitterly cold winter morning with a call from Northern Constabulary. Normally when the police ring me it is to talk about a case, but this time the business was personal: my father had been found outside a care home in his joggers and T-shirt at five o'clock in the morning, in temperatures of –10°C. The police had taken him into the care home, assuming this was where he'd come from, only to be told he wasn't 'one of theirs'. They warmed him up and fed him biscuits and coffee while they talked to him to try to find out who he was and where he lived. He clearly had enough presence of mind to take them to his house, where they found the door wide open and, in the kitchen, a list of telephone numbers pinned up on the kitchen wall by my ever-practical mother. That A9, known locally as the 'road to hell', doesn't get any shorter, or the speed cameras any fewer, no matter how many times you go up and down it.

It was obvious that it was not safe for him to continue living on his own. The strong, stubborn man of my youth would have to be taken into care.

I remember, when I was very young, my mother's aunt Lena going 'doolally' or 'lally doo tap', as my father described it. Her advanced dementia had taken her to Craig Dunain Hospital, where he would visit her every week. She was totally non-responsive and didn't recognise him, yet this undemonstrative man would sit with her and talk at her for hours at a time, while she rocked backwards and forwards, gently rubbing her finger and thumb together incessantly. I asked him one day why he bothered and his answer, in which I heard echoes of his own mother, shocked me so much that I never forgot it. 'How do we know she doesn't hear us?' he said. 'How do we know that she isn't just locked inside her head and simply can't communicate? How do we know she isn't lonely and scared?' He wasn't prepared to take that risk and so he visited to talk to her and keep her company at a time when my mother found Lena's condition too upsetting to witness. It was a side of my father that took me completely by surprise. And so, when his own dementia came calling, I never once assumed that he wasn't there, locked inside his own head, afraid and alone.

My father lived to almost eighty-five, and for the remainder of his years we watched as, very slowly, bit by bit, this six-foot giant of a man, his bristling military moustache, his bandy legs, his barrel chest and the voice that could stop traffic dwindled until, finally, he all but disappeared. With the ravages of the disease, some early distressing bouts of rage eventually gave way to placidity. We moved him to a care home five minutes' down the road from us in Stonehaven, where my little family became his sole companions for nearly two years. It was too far for his old friends to travel, and he no longer remembered them anyway. Grace, our fledgling nurse, saw him the most as she took a part-time job in the home. We wondered if this work experience would put her off her choice of career but it seemed only to fuel her determination.

We had a good couple of summers and Christmases with him and, from a perhaps selfish point of view, those years gave us an opportunity to spend time with him that we would all cherish for ever. Time to chat, to listen to music with him, to sing and go out with the wheelchair he needed after falling and breaking his hip.

I would sit with him in the sunshine and hold his hand—a tactile demonstration of affection I could never have contemplated as a child. He loved the warmth of the sun and when we took him out into the garden he would turn his face up to it and bask contentedly like a cat. He clearly still derived tremendous pleasure from these activities, and from his Maltesers, ice cream and the odd little dram. The moustache would twitch after the first sip and rosy spots would appear on his cheeks. He did not seem to be in pain, he did not seem distressed and there was no doubt in my mind that he knew who we were because his face would light up when we came into the room.

Yet how the man he had been would have detested his dependency on others, including me. The nurses at his care home were genuinely fond of him because he was never any trouble and always had a twinkle in his eye and a smile for them, which was such a solace. Although none of us was 'happy' for him, he was safe, he was well cared for, he was loved, he was warm and clean, he was pain-free and he led a calm and peaceful existence. That said, it was still a soulless environment—functional, comfortable enough, but always clinical, never homely. My father would have called it 'God's waiting room'.

In the last year of his life he forgot how to walk and then how to speak. And then, slowly, he started to shut down. One day, as if he had decided that enough was enough, he stopped eating. Soon he stopped drinking. He metaphorically turned his face to the wall and just waited for the end to come. Maybe he was even inviting it, I don't know. I had power of attorney for his health and gave the same instructions as I had given my mother's doctors: he was not to be resuscitated, not to be put on a drip, just made comfortable, kept free of pain and allowed to die when he was ready.

When death did come for him it was not with violence but calmly, quietly and patiently, at a pace of which he would have approved and that perhaps he drove. Realising that time was short, Tom, Beth, Grace, Anna and I visited him on what turned out to be his last day. It seemed that, to all intents and purposes, he had simply decided to switch himself off. He lay curled on his side on the bed, not registering that there was anyone in the room with him. If he could hear us, he heard chatter, laughter and his favourite music—'Highland Cathedral', performed by our girls' school pipe band—on the CD player. He remained utterly motionless and non-responsive. He wouldn't take any fluid, and the skin on his enormous hands, like bear paws, though warm, was as dry as fine paper.

When it was time for us to go for the night, I told him, as I had told my dying mother, that we were leaving, but would be back in the morning and he'd better still be there. Old habits do die hard. An unmistakable look of terror crossed his face and flashed out from those expressive, black eyes. I was stunned. My father had been virtually non-communicative for months. Beth gasped. She had seen what I had seen. I had not imagined it. 'Mum, I don't think you are going anywhere,' she said. He had been right, all those years ago, to question our assumptions about Aunt Lena. He *was* still in there, locked in his own silent world, unable or unwilling to communicate. Now, when it really mattered to him, he had found the strength to send out an SOS in the only way he could. He knew what was coming and he did not want to be alone.

As a child I had promised my grandmother I would be there for him when the time came, and evidently this was it. I reassured him that I was going home just to shower and change, and would be back within the hour. When I returned, Tom, Grace and Anna left; Beth chose to stay with me and her grandfather.

I don't think my father was afraid to die, just anxious about dying alone. His mother had understood him so well. Beth and I sat in the dimly lit bedroom and talked and laughed and sang and cried. He made no response but we held those huge hands and he was never alone for a moment. If one of us left the room for a toilet break or to fetch coffee, the other stayed. He did not move a muscle. His hand never clasped around mine, his eyes never opened. There was absolutely no doubt that this night was going to be his last—everybody knew it, him included—but the atmosphere was one of peace.

In those wee small hours when the ghosts of a life visit for the last time, his breathing became shallow. I told him it was OK to go, that we were with him, that he was not on his own. His breathing slowed, slowed further, deepened and then stopped. I thought it was all over but then he took a few more shallow breaths. There was a short spell of agonal breathing—gasping, basically—before the sound of the death rattle, caused by mucus and fluid collecting in the back of the throat, where it can no longer be coughed away. Finally, the last gasp, nothing more than a brainstem reflex. Within a matter of seconds, when I saw the foam from his lungs appearing on his lips and nose, which meant there was no air left in them, I knew he was dead. It was as simple as that. No fuss, no distress, no pain, no hurry—a gradual giving out of power.

The huge physical and spiritual presence that had been a cornerstone of my life had slipped away from this world in what seemed like the flick of a light switch. A smaller, thinner shell remained but the presence had left the room. It was a strange sensation: I felt no attachment to this shell, because it was not him. My father was not his body, he was something very much more than that.

We opened the window to let his soul fly. If his mother was there to meet him, as she'd promised, I had no sense of it. I was not, of course surprised, but perhaps I was a little disappointed. Then we cried for a bit before settling ourselves down and getting on with what needed to be done. We fetched the nurse, who checked his pulse (we'd done that) and his breathing (we'd done that, too) and she confirmed a time of death—a good ten minutes later than when it actually happened, but that hardly mattered.

My father had technically died of old age. In days gone by, the cause might have been expressed on his death certificate in more poetic language but in the more banal medical vocabulary of today it was attributed, like so many others among the elderly, to acute stroke, cerebrovascular disease and dementia. I was there: he had no acute stroke. He probably did have some form of CVD (we can all expect to with age), but last time I read about it, dementia doesn't kill you. It was simply his time, and he had chosen it.

I had, though, felt throughout his illness that Alzheimer's was a cruel route to death. The long, drawn-out period he spent dying was hugely distressing for us all and, I suspect, for him, in those moments of clarity that probably visited him in the night when he was on his own. We felt we had begun the grieving process at least two years before he died, when we began to lose the man we knew. But in the end, he had a 'good death', just one that took too long to arrive. His time was up, he turned his face to the wall and he died peacefully with those who loved him by his side. Can it really get much better than that?

CHAPTER 5

Ashes to ashes

'The measure of life is not its duration, but its donation'

Peter Marshall pastor (1902–49)

My grandmother, Margaret Gunn, in Inverness in 1974.

If the way we support and comfort the dying is broadly similar across countries, cultures and belief systems, the same cannot be said for funerals. But whether we are talking about the Buddhist sky burials of Tibet—in which the body is returned to the earth by being chopped up into pieces and left on a mountaintop—the famous brightly coloured, noisy jazz processions of New Orleans or the more sombre occasions traditional to the UK, they all provide a reassuring template for mourners to follow at a time of raw emotion. These ceremonies are very important, not only in enabling families and communities to commemorate the lives of the deceased and bid them a public farewell, but also in bringing some solace to the bereaved by giving them a framework within which to ritualise their grief, whether that involves expressing it or masking it.

The stark truth is, of course, that grief never dies. The American counsellor Lois Tonkin reminds us that loss isn't something we 'get over', and it doesn't necessarily lessen, either. It remains at the core of us and we just expand our lives around it, burying it deeper from the surface. So with time it may become more distant, more compartmentalised and therefore easier to manage, but it does not go away.

The bereavement theory developed in the 1990s by the Dutch academics Margaret Stroebe and Henk Schut suggests that grief works in two primary ways and we oscillate between them. Their 'dual process' model of grief defines these as 'loss-oriented' stressors, where we are focused on our pain, and 'restoration-oriented' coping mechanisms involving activities that distract us from it for a while. All we can hope for is that the periods of paralysing, overwhelming grief become less frequent. But living with loss is personal to all of us and has no predetermined path or timeline.

The funeral of a loved one is just an early step on that path. In the UK, for the most part these rituals used to be rooted in the Christian Church of one denomination or another, but as the country has become more multicultural and secular, so have the ways in which we mark a death. In general, as a nation we are becoming less religious, and while our hospital beds fill up with those desperate for cures, our church pews are emptying of those who rely on their faith. Where in the past we might have accepted a terminal prognosis and turned to a church to ensure the health of our souls, now we are more likely to trawl the internet in search of every last vestige of temporal hope that might keep us alive for just a little bit longer.

The solemnity, propriety and ceremony surrounding death are fading as it grows more secularised. Gone are the weeks of professional mourning of bygone years, the mourning jewellery worn from the Middle Ages until Victorian times (I have a great collection of this, by the way), the doffing of caps as a funeral cortege passes, the *memento mori* which, I must admit, I've always found a bit creepy. Going are the hymns of old, to make way for Frank Sinatra or James Blunt. Our anatomy department recently had a gentleman inquire if we could possibly embalm his body as he wanted to be buried sitting on his Harley-Davidson and couldn't think of any other way that he could muster sufficient corporeal rigidity. Utterly bonkers, if ingenious—we had to say no.

I was definitely born in the wrong century. I prefer a decent send-off like the traditional funeral processions you still see in London's East End, the shining black carriages drawn by black horses bedecked with plumes, led by a top-hatted funeral director who sets a proper, respectful walking pace. They are magnificently spine-tingling in their pomp and ceremony.

I love a good graveyard, too. They are wonderfully peaceful and welcoming places, especially those in town centres, where their prime position reflects their importance to the community in times gone by. My grandmother and I would take a picnic up to the top of Tomnahurich Cemetery (always referred to by my father as the 'dead centre of Inverness') when we visited her husband in summer and my husband Tom used to run up and down its steep tracks when he was in rugby training. So many cemeteries have been abandoned and become derelict, perhaps paving the way for a future in which we create electronic graves, where family and friends can post memorial images online. Not quite the same thing in my book.

As we grow older, we attend increasing numbers of funerals and take more notice of changes and trends as the old customs are swept aside to make way for how we think things should be done today. If I regret the disappearance of some of our longstanding conventions, I would acknowledge that in many respects the freedom we now have to mark a death by choreographing a farewell that more specifically reflects the identity, personality and beliefs of the departed is a positive development. And although mourning rituals are less protracted and less public, the grief is still as real. If the purpose of consoling the bereaved while honouring the dead is fulfilled, then who are others to stipulate how this should be done? Equally, tradition still matters for as long as there are those who gain comfort from it.

There is so much to do before a funeral can be held that sometimes you wonder whether the whole process has been expressly designed to keep you busy to distract you from your grief. Alongside registering the death, arranging the funeral director, obtaining copies of the death certificate and putting a notice in the newspaper, there are so many decisions to be made. Both my parents' funerals took place at a crematorium chapel, which meant choosing flowers and hymns and writing a piece for the minister to deliver. Did we need funeral cars, and if so, how many? There was a casket to be selected (my father would have commented of his that he had burned better, which was ironic as that was exactly what we were going to do with it), a venue and catering arrangements to be decided upon for refreshments afterwards and we needed to make sure the right people were informed. In Scotland, where the interval between a death and a burial is short, the fevered activity necessary to get everything done in time brings out the best and the worst in people. There are inevitably moments that will go down in family lore.

My father had been a church organist for many years and I knew what he would have liked to be said and sung at his funeral, and what he definitely wouldn't. Yet however desperate I was to do him proud, I couldn't escape the feeling that it was ridiculous to be still thinking about his preferences when he was the one person not going to be there that day to care one way or another.

On Saturday evenings, my father used to go to the kirk to practise before the Sunday services. Sometimes I would join him and just sit in the front pew listening to his lovely touch on the church organ. He would often choose to play Glenn Miller's 'In the Mood'. It was odd to hear that big-band tune blaring out in the empty kirk, but I loved it. On Sundays, as a young girl, it was my job to go to church with my father, position myself in the second pew from the front, directly opposite the organ, and keep an eye on the hymnbook during the singing. When we reached the last verse of a hymn I had to remember to put my hand on the back of the pew in front of me—the agreed signal for him to stop at the end of the verse. There were a few occasions when I forgot and Father would be happily playing another verse that didn't exist. I usually got a row that day.

Father hated it when the congregation didn't sing up. So it got to me that, at his own funeral, the mourners were mumbling the hymns. I couldn't bear looking at the poor organist in the corner knowing how annoyed my father would have been. I did the unthinkable. I moved forward, threw up my hands and shouted for everyone to stop—yes, in the middle of the service. I told them how my father felt about playing the organ when people didn't sing from their hearts, and asked if they could please give it some welly, just for him. My daughters were horrified and most of the rest of the congregation thought I had lost my mind. But I do like an occasion to be memorable.

I'd had no difficulty choosing the tune to be played as people were leaving. What else could it have been but 'In the Mood'? Or 'In the Nude', as my father used to call it.

Both my father and my mother had stated very clearly that they wanted their remains to be buried, but they were not bothered whether those remains were their bodies or their ashes. Of course, there is a third option, but neither of my parents wished to leave their bodies to anatomy and I didn't ever feel it was my place to try to persuade them otherwise.

So far, so sensible. The lunacy lay in the location of their burial sites. Mother wished to be laid to rest with Uncle Willie and Teenie at the bottom of Tomnahurich Cemetery and my father wanted to be with his parents at the top. We had suggested that maybe they would like to be buried together, but good Scottish pragmatism (or, in my father's case, short arms and long pockets) took over. There was one empty space in the lair at the bottom of the hill and another at the top, both bought and paid for. Why waste money on a new one? They both felt that when they were dead, they were dead and they didn't really care about where they were buried as long as it was done properly. They may have been traditionalists, but they were also practical and unsentimental. Father always promised he would wave to Mother from the top of the hill and she always retorted that she'd ignore him.

And so my father was cremated, and for about a year he sat in a lovely, well-crafted box, of which even he would have approved, on our hallway table until we could get the whole family together for his interment. I didn't feel there was any hurry. He was dead and he wasn't going anywhere. Even our cleaners, after the initial shock, got used to him being there and became rather fond of him. They would say good morning to him when they came in the front door and give his brass plaque a bit of a dust down. They were quite sorry when he eventually left. People don't have to be alive to make their presence felt.

On Christmas Day we decided that Grandad should join us for lunch and so we placed his box at the end of the dining table. While that might sound weird to some, it felt somehow normal for us to have him with us, a Santa hat perched on top of his box. We raised a glass to all those absent who meant the world to us, and to him—the last member of his generation to go.

This generational shift in the family had an impact on Anna, the youngest, as she got to grips with the fact that her father and I were now the oldest generation in our family, and that she and her sisters had been promoted to second in command. So she found my father's death difficult not only because she adored him, but because she was terrified by the thought of whose turn it might be next.

When the time was finally right for my father to be buried, we gave that honour to my sister's son, in whose life he had been an important influence. Barry displayed great dignity as he carried Grandad from the boot of the car to the hole in the ground and, with great solemnity, carefully planted him there. Anna decided that Grandad would need a dram to help him on his way, and poured a good shot of Macallan over his box once he was installed. He would probably have considered that a waste—a view evidently shared by the ever-watchful gravedigger lurking in the background.

Whatever we believe happens to the soul, or the essence of

a person, after death, the bereaved usually feel a visceral need for a specific place they can visit, or picture in their mind's eye, where the mortal remains of their loved one reside. For some this will be a grave; for others a wider landscape where cremated ashes have been scattered, generally a location that meant something to the deceased in life. Many people choose to keep ashes with them, as we did with my father's, for a while, sometimes permanently. Some even bring them along on days out the person might have enjoyed in life, or to places they never managed to get to see. I know of someone who took their mother's ashes to New York for a weekend because she had always wanted to visit Central Park.

Cremation, first introduced in the UK early in the twentieth century, is now the choice of a majority of people and its popularity is evidenced by the number of imaginative things you can now do with somebody's ashes. They can be fired into space or deposited in water to create a marine reef; you can have them incorporated into glass and made into jewellery, paperweights or vases. They can be put into shotgun cartridges, turned into fish bait or added to fireworks to ensure your send-off goes with a bang, or even compressed to create teeny little diamonds.

When there is no chosen 'resting place', and a proper funeral is not possible, it is hard on families—indeed, it is one of the lasting agonies suffered by the relatives of likely murder victims, or those killed in disasters, whose bodies are never found. So forgoing these rituals at the time when their loss is at its sharpest is a big sacrifice to ask of the families of those who, like Henry, the man who taught me from the dissecting table, decide to donate their bodies to anatomy or to other scientific research. I understand completely how relatives may be left feeling they have no 'closure'. A body bequeathed to science can be retained by law for three years—a long time for a family to wait for the ashes of their loved one to be returned to them. But in the case of these donors, we hope the certainty that the firm wish of the deceased is being fulfilled brings some comfort.

The decision to leave your body to medical, dental and scientific research and education is not one to be taken lightly. The reasons why some people opt for this route are many and varied, but they are mostly altruistic, arising from a genuine desire to play a part in advances that could save lives or alleviate suffering. Some bequeathers simply believe that 'dead is dead', and their remains might as well be put to good use as destroyed or allowed to rot. As one sassy elderly lady once said to me, hands on hips: 'Young lady, this is just too darn good to burn.' For others, the reason might be starkly practical. When you consider that the average cost of a funeral and burial in London is about £7,000, and just over £4,000 across the rest of the UK, you can see the economic appeal. But we do not judge anybody's motives. It is a personal choice and our job is just to help people to make it happen.

In our anatomy department at Dundee University we have a dedicated bequeathal manager, Viv, who takes calls every day from people inquiring about donating their bodies. An anatomy department is one place where you can be confident that a conversation about death will have no uncomfortable silences, platitudes or condescension. Some prospective donors ask to come and visit us to talk about the practicalities, or to look at our Book of Remembrance. Others just want to put arrangements in place with as little engagement as possible with the process. In these circumstances, Viv pops the necessary forms in the post to them—although I have known her get in her car and deliver the paperwork herself when she is dealing with someone too infirm to visit but who she feels needs the personal contact.

The bequeathers sign the forms in front of a witness (not Viv—that would be improper), send one back to the anatomy department and lodge the other with their will at their solicitor's office. And that's it. We do, though, actively encourage them to speak frankly about their wishes to their families and carers, so that when the day arrives nobody is caught by surprise and delays can be kept to a minimum.

People who choose to donate are not looking for cloying niceness or obsequiousness, they just want warmth, reassurance, trust and honesty. When they ring Viv they have come to the right person. I marvel when I hear her on the phone. A kind woman with a wicked sense of humour, her objective is to answer truthfully, directly and with humanity any questions put to her and never to placate with vague words of comfort. She has regulars who phone her up just for a chat, to let her know that they are still alive and regale her with details of their latest ailments. They view her as a friend—the person who will be there for their family when the dreaded day arrives. And she always is.

When the call finally comes from a son or daughter, husband or wife, Viv steers them gently but firmly through everything that needs to be done to get the body to our department as swiftly as possible. This can be a challenging time for families. They might not understand or agree with the decision of the person they love and often feel confused by the inherent lengthy postponement of the usual funeral ceremonies. We do the best we can to assist in the fulfilment of a donor's bequest but, since we have no desire to cause additional pain to relatives, at times strong family objections may overrule the wishes of the deceased.

As well as consenting to their bodies being kept by us for up to three years, bequeathers can opt to give permission for some body parts to be retained for longer, for photographs to be taken for educational purposes and for their remains to be used by another department in Scotland if ours is unable to accept them. This is a lot for someone to take in when their mum has just died, which is why we advise all our donors to speak openly and honestly with their families about their decision.

Viv's is the most important, delicate and compassionate public-relations job at the university, and she accomplishes it flawlessly at a time when family grief is at its most acute. She was recently awarded an MBE for her services to anatomy bequeathal in Scotland—not for her 'services to dead bodies', as some crass journalist put it. I am so proud of her and of the work she does.

◊

Our donors come from all walks of life. We have postmen and professors, grandads and great-great grannies, saints and sinners. The youngest age at which we can accept a donor in Scotland is twelve but the vast majority are over sixty. Our oldest to date was 105. The life lived is of little consequence to us and we accept almost everyone. There are one or two instances in which we might have to refuse a bequest but they are rare. If the coroner or procurator fiscal has required a postmortem, we cannot accept the body as it will have been disrupted by the examination. If the deceased had such extensive cancer metastases that little normal anatomy remains, we may decline and in the past we have occasionally had to turn down the morbidly obese for the simple practical reason that our equipment was not able to cope with lifting them.

About 80 per cent of our donors are local to the university region and we take great pride in the relationship we have with the Tayside community. We now have several generations of Dundee families who have 'gone to the university'. Their names are recorded in our Book of Remembrance. This is not just a memorial to the bequeathers but a daily reminder to our students of how fortunate they are to benefit from the gift of so many people who have asked only one thing in return: that they learn. The book is displayed at the top of the stairs to the department so that every single student will see it every time they enter the dissecting room.

One donor who exemplifies our relationship with the local community is an elderly man I'll call Arthur. He is a delight: he comes to all our university events, whether the lecture on offer is on forensic science or creative writing. He has an active mind, is hungry for experiences and remains a deep thinker who ponders on his legacy but not his mortality. He isn't religious and he sees the merit of, in his words, 'recycling' his remains for the general good and not spending an unnecessary fortune on a 'wasted funeral ceremony'.

Arthur is, however, distinctive in having planned his own exit strategy from this world. He is adamant that he does not want to be dependent on others should he become infirm or incapable in his advancing age. When he has had enough, he wishes to take responsibility for his own death and end his life by his own hand. He does not want neighbours or friends to find him undergoing the indignities of dying. He is in full charge of his mental faculties, his mind is made up and no debate has ever led to him modifying his ideas—believe me, I have tried enough times. Having researched the matter exhaustively, Arthur has now chosen the way he means to die. He tells me he has purchased equipment from the internet that will allow him to go peacefully, will not cause any disruption to his body and will leave him fully in control of his actions and his decision until the very last moment.

These are not thought processes that many of us follow in such detail through to the conclusion Arthur sees as natural, although we will probably all understand them in an abstract way, and some of us will relate to them. Assisted suicide and voluntary euthanasia remain illegal in the UK. Government bills come and go and I believe that eventually one will succeed in permitting us to make the choice, in circumstances where we wish to exercise it, as to how and when we end our lives.

I think that one day we will be able to make this mature decision without pressure from the authorities, and with proper legislative controls in place, so that those of us who wish to have some control over our death are not forced to come up with the funds needed to die in a foreign country to achieve it or to take more drastic measures.

Suicide tourism is an expensive business and the decision to embark on it is often made earlier than is necessary because of fears that delaying too long may result in the person becoming too ill to undertake the journey. In making sure this doesn't happen, they may well be depriving themselves and their families of a few more precious moments and experiences together before reaching the point where no quality of life at all is possible.

Assisted suicide (or assisted dying) is legal in Canada, the Netherlands, Luxembourg, Switzerland and parts of the USA. In Colombia, the Netherlands, Belgium and Canada, voluntary euthanasia is also within the law. The difference between the two practices lies in the degree of involvement of a second party. If a patient asks a physician to end his life, perhaps with a lethal injection, and the physician complies, this is defined as voluntary euthanasia. If the physician prescribes lethal drugs for the patient to self-administer, it is assisted suicide.

In America, assisted suicide is legal only for those diagnosed as both terminally ill and mentally competent who die in Oregon, Montana, Washington, Vermont or California. Oregon was the first US state to legalise assisted dying with its Death with Dignity Act of 1994. Medication can be prescribed by a physician and self-administered only after two doctors have confirmed that the patient is likely to have less than six months to live. Strict safeguards have ensured that there have never been any proven cases of abuse. The authorised drug is a mix of phenobarbital, chloral hydrate, morphine sulphate and ethanol and costs between $500 and $700. Approximately 64 per cent of patients who request the medication will take it, usually in their own residence. The fact that the remaining 36 per cent, quite a high number, decide not to use the drug illustrates that people understand the nature of choice. Perhaps simply knowing that the drug is there if they want it is enough to reassure the terminally ill that control of their own life and death resides in their own hands.

In UK hospitals, the terminally ill may have little personal control over their last moments and their relatives must rely on medical staff to ensure that their dying and death is as pain-free as possible. Physicians may employ continuous morphine sedation and food and water may be withdrawn, which can result in death occurring relatively swiftly, as happened with my mother.

The British Medical Association regularly votes against assisted dying, perhaps understandably fearing that it would have a detrimental effect on society's trust in doctors. Yet in a recent European survey, the country with the highest level of trust in doctors was the Netherlands, where assisted death is legal. It seems that being given a choice may increase trust rather than diminish it.

The arguments for and against legalising assisted death are well rehearsed. Supporters maintain that just as we have a right to life, we should have a right to a dignified, humane and pain-free death at a time of our choosing. Opponents express concerns about the dangers of any legislation being abused, about potential societal pressures on the elderly or infirm not to 'become a burden', about illness or disability being perceived as a justification for ending a life. Some disagree for religious reasons, believing that only the Creator has the right to decide when we die. The voices of detractors often drown out the views of the unfortunate people suffering agonies they consider intolerable and inhumane who are desperate to have the option of assisted dying available to them. It is not illegal for them to end their lives, but to comply with the law it must be done without assistance, which means that often the only options at their disposal are traumatic or violent.

Whatever the viewpoint, choosing when to die should, in my opinion, be a personal matter, not a decision controlled by the state. Perhaps the adoption of a less pessimistic and mistrustful approach to the wishes of those who seek the freedom to decide the manner and timing of their death can be seen as an indication of a responsible society. It is probably no coincidence that those countries and states where assisted dying is legal usually have a higher investment in palliative care and are generally more open about death and end-of-life options. I for one would prefer to be part of a society that allows people to have greater control over their own lives and deaths.

I respect Arthur and his determination to die on his own terms, and I share his resentment that at present, society forces him to consider undertaking this alone because it is unable, or unwilling, to find the flexibility of legislation that would allow him the dignified exit he so desires. His determination to bequeath his remains to an anatomy department mercifully rules out the most violent means in his case: as he wants to avoid a postmortem, he doesn't want to 'disrupt his body'.

He has spoken to us of avoiding Christmas and the New Year, when the university is closed, asking which days are likely to be most convenient for anatomy departments. I feel a heightened sense of anxiety when he talks this way, but I also know that there is nothing I can do to dissuade him, because we have had these conversations many, many times. I will not aid him but I cannot stop him—that is not my right, nor is it an option he gives me. I count it a privilege that he feels he can talk to me and I will not interfere, just allow him to rehearse his rhetoric, testing how comfortable and reasonable it sounds, to both himself and others.

Arthur was deeply saddened when, even after taking all

this into consideration, he approached another anatomy department to seek their views on his plan, only to be told that his body would not be accepted by them if he were to commit suicide. He found it difficult to reconcile this stance with his understandable wish for a 'good death' and his genuine ambition to assist in the education of others.

He has thought of just about everything. He has given me a code word only he and I know, which he says he will leave on the answering machine in my office over a weekend so that it is waiting for me on the Monday morning. This will be the signal for me to alert the necessary authorities so that they can begin to arrange for his wishes to be followed. He will not tell me in advance when he is going to die, to protect me from any suggestion of involvement and also because he doesn't want me to try to stop him. It is, in an odd way, a kindness, but it has led to me developing a very healthy antipathy to the flashing red message light on my phone, especially on Monday mornings. So far it has never been an indicator of a message from Arthur and I hope it never will be. While I must acknowledge the possibility that he will one day execute his plan, my hope is, of course, that when the time comes he will experience a peaceful, swift and natural end that both accommodates his wishes and allays society's current fears and restrictions. In case I am on holiday or not in the office, Viv has also been briefed. Arthur has us both totally wrapped around his little finger.

It is hard to put into words how grateful I feel to Arthur for his strong support for bequeathal and anatomical education, and for sharing his most personal wishes with me, but I also feel a tremendous weight of responsibility for ensuring that they are respected while all legal requirements are upheld. The moral conundrums are heavier still. This is where the real wrestling is done, late at night when he pops into my mind and I wonder what he is doing. Is he lonely? Is he well? Is he scared? Is he putting together the different parts of his exit equipment? Can I stop him? *Should* I stop him? Although he has my phone number, I do not have his. I have no idea when he intends do this, if he ever does, and by the time he has done it, it will be too late for me to intervene. So all I can realistically do is to keep talking to him.

I am not sure if I want him to change his mind, if it means leaving himself at risk of the kind of death he so emphatically rejects, but I feel that if I keep asking the questions I am at least prompting him to continually reassess his decision. He gets quite cross with me at times for my persistent poking and prying. I tell him my questions are coming 'from the loving place', at which he usually grimaces dismissively and says, 'It's not a very nice place, that loving place.'

He has a habit of throwing in some curveball questions himself, it must be said, outlining theoretical situations that make you pause and take stock. He does so with a devilish twinkle in his eye. Quite some time ago now he asked if he could look at our dissecting room and watch some dissection. I gasped. Never before had we had a bequeather requesting to see what goes on in our dissecting room. But why was I knocked off centre? Who are we protecting? You can buy a ticket to walk around the Body Worlds exhibition and see a whole array of dissected humans in different poses. You can go to a surgeons' museum and peer into glass cases containing spine-chilling pathologies and anomalies of all sorts excised from human bodies, examine the gruesome and the grim embalmed in formalin and mounted in a glass pot. On the internet you can call up all manner of images associated with the dissection of human cadavers. You can pop into a bookshop and pick up an atlas of human dissection, or watch the procedure on television. Arthur seemed to have no qualms about seeing the dissecting room; I, on the other hand, was inexplicably and massively conflicted. Was it just too personal for me to handle, or too big a responsibility?

One day Arthur will be a cadaver in someone's dissecting room, if he has his way, and I don't doubt for a moment that he will. As he aspires to be a cadaver, it was perfectly reasonable that he might want to see what he may look like on the inside, and the type of environment where he might spend several years. When prospective students come to visit the university, they are permitted to view the dissecting room, so why not prospective donors, who are, after all, the other half of this symbiotic relationship? Maybe, thinking back to my own first experiences in a dissecting room, I was afraid it might frighten or disturb him. There was just no way of knowing whether it was more likely to be an unmitigated disaster for him or a tremendous success for his peace of mind.

I tried to brush away his request with a glib remark, but he wasn't going to let me get away with that. I was told politely but firmly that he wanted to do this with me because he knew and trusted me, but if I wasn't comfortable with it, he completely understood. He would go to another department and ask them. Such a little blackmailer! I heard somebody somewhere say, in my voice, that I would check with the authorities that it would be OK, and so it seemed I had agreed. Reluctantly. I have always been unable to say no to Arthur and I am not quite sure why. Maybe it is because I like him so much and am so proud of the work undertaken in my department by staff who are utterly committed to donors, family, students and education. If our 'silent teachers' are 'teaching', then they are staff. Perhaps, at a stretch, I could think of Arthur as a prospective future member of an anatomy teaching team. I knew that if I ran that one past him, he would chortle with disdain and probably accuse me of exploiting him as cheap labour.

I checked with Her Majesty's inspector of anatomy, and he had no problems with this arrangement as long as it was a controlled visit. So, on the appointed day, Arthur and I met in my office and talked again about bequeathal and what it meant to him, to me and to our students. We discussed his plans for his death, I did my best to get my views across and, as always, they fell on conveniently deaf ears. I explained the embalming process and he questioned me on the chemical reactions that occur at the cellular level. He asked about smell, touch and sight. We looked through some textbooks and he commented that the muscle tissue didn't look as red as he'd expected. He told me he'd been imagining it as a similar colour to the meat you'd see in a butcher's shop rather than the pinky grey it actually is. It was good for him to see these pictures to prepare him for what he would encounter in the dissecting room.

We chatted about the skeleton that hangs in the corner of my office and the colour-coded markings painted on it to identify where different muscles originate and insert. We handled the skulls that sit on my bookshelf and discussed how the bones grow and how they break. Over a cup of tea, we talked about life, death and learning. I let him set the pace.

When he was ready, we made our way up from my office to our museum. Arthur was already getting on by then and very bent, and the steps were difficult for him, but he managed, gripping the rail with one hand and his walking stick in the other. We stopped for a moment and I pointed out our Book of Remembrance in its glass case at the top of the stairs. Arthur remarked on the number of people who donate their bodies to us and theorised about their motives. We spoke of our memorial service in May of each year and he asked about the ages of the youngest and oldest donors in the dissecting room at that time. Did we have more men or more women? I answered all his questions honestly and openly.

As we moved along the corridor we passed the amazing work done by our talented medical and forensic art students and talked about the ancient relationship between anatomy and art, with particular reference to the glorious Dutch masters, who had a morbid fascination with anatomical dissection.

Our museum is in a bright room, furnished with rows of long, white tables where our students study and compare prosected specimens with the illustrations in their textbooks. Arthur took a seat at one of these tables. I showed him the sagittal, coronal and horizontal sections of human bodies, displayed in heavy, Perspex pots, which allow us to teach anatomy in slices that relate to the images produced by CT and MRI scans. I hauled one pot over to the table where he sat, informing him that this was a horizontal section through the chest region of a man. 'How do you know?' he asked. I pointed to the hairs poking out of the skin, and we both chuckled.

I indicated the position of the heart, the lungs, the major blood vessels, the oesophagus and the bones of the ribs and vertebral column. Arthur was utterly intrigued. He expressed surprise at the diminutive size of the spinal cord, which carries all motor and sensory information around our body, and the oesophagus, remarking that in future he would make sure he ate his food in smaller mouthfuls. He commented that seeing how delicate some of these structures were made him realise just how fragile life is. He looked at the coronary vessels in the heart, and the widow-maker artery (the anterior interventricular branch of the left coronary artery), and asked me to identify the chambers of the heart that were visible. He was amused by the chordae tendinae, colloquially referred to as the heart strings, which sound so romantic. In reality, they looked to him, he said, like miniature guy ropes holding down a Lilliputian tent. He asked how old the specimen was and how long it would survive.

I was easy in my mind that this elderly gentleman was completely comfortable with what he was seeing and discussing. I detected no apprehension, except perhaps from myself. There was no fear in his rheumy eyes, no tremor in his voice and no shake in his hands. Time for the big one. I left Arthur perusing the pot for a moment and popped into the dissecting room, a light, open space full, as usual during working hours, of chatter and students going about the normal business of an anatomy department. I scanned the room for a table of more mature students. Finding a group that fitted the bill, I told them about Arthur and asked if they would be willing to talk to him. It was clear that they were unnerved by the prospect of having a conversation about dissection with a trainee cadaver—especially while they were standing over someone else's body, scalpels and forceps in hand, in the middle of opening up the shoulder joint. But they mulled it over, discussed it among themselves and decided they were up for it. A spokesperson was elected.

I don't know who was more scared—the students, Arthur or me. I still had no clue how this was going to turn out. Would it be a colossal mistake? Arthur got to his feet very slowly and walked with me into the dissecting room. You could have heard a pin drop. The cheerful banter of a few moments before was gone, replaced in an instant by respectful silence and diligent attention to work. It is amazing how, as if on some unspoken command, the entire atmosphere in a room can change in a split second. There is a sudden collective awareness that someone outside the close-knit team is present and a uniform modification of behaviour ensues. We see this all the time in mortuaries, where there is an unwritten rule that when a stranger enters, you adjust your conduct and demeanour until you have a good idea of who they are and what they are doing there. All the students in the dissecting room did this without being given any warning or instruction. I was so proud of them all.

Arthur approached the table a little hesitantly. The lead student introduced himself and joked nervously that perhaps shaking hands wasn't appropriate given the work on which they were engaged. The other students around the table then introduced themselves. They were so pale and nervous that I thought one or two of them might pass out. Arthur pointed at the table and asked, 'What is that? Why have you cut it that way?' I stepped back and watched the most amazing miracle unfold in front of my eyes: Arthur and the students, far from being separated by death, becoming united by it through the glorious world of anatomy.

The chatter level in the room began to rise again as he was accepted into the circle. Arthur talked to his dissection team for a good fifteen minutes or more. Once or twice I heard relaxed laughter at something he said. Feeling that a quarter of an hour was long enough for all of them, and for Arthur to be on his feet, I moved in to shepherd him away. He thanked the students for their professionalism and they, in turn, thanked him for the priceless gift he was planning to make. I sensed a genuine reluctance on both sides to end the conversation. Mind you, I also noticed the students' collective sigh of relief when Arthur turned and started to slowly walk away. They had been really afraid of offending him or upsetting him. But they understood the importance of what they had done for him and, indeed, what he had done for them, and would do for future students.

For Arthur, it was back to my office for that obligatory restorer of comfort—more tea—and a little chat. He was enthused, animated and more determined than ever to see through his plans to donate. His only regret, he said, was that he would be on the wrong end of the scalpel. He had found his encounter with the dissection process so fascinating that I wonder if, had his life followed a different path, he might have made a great anatomist himself.

It was an intense experience and it had an incredible impact on everyone involved. So would I ever do it again? Good grief, no.

CHAPTER 6

Dem bones

'There is something about a closet that makes a skeleton terribly restless'

Wilson Mizner playwright, entrepreneur and raconteur (1876–1933)

A facial reconstruction of Rosemarkie Man.

At what point does your death cease to matter personally to someone somewhere? In his poem 'So Many Lengths of Time', Brian Patten suggests that 'a man lives for as long as we carry him inside us', and that certainly strikes a chord with me. So often, as I grow older, I open my mouth and my father's sayings fall out. We cannot die as long as there are people on earth who remember us.

By that yardstick, we have a potential 'lifespan', or should that be 'deathspan', of probably no more than four generations, though echoes of us can live on for longer, in the memories of relatives, family stories, photographs, film and other records. In my family, my generation is the last to remember my grandparents, and my children are the youngest to remember my parents as my grandchildren never knew them. It saddens me that when I die, so, at last, will my grandmother. Yet I find it apt and comforting that we will die together, I in my body and she in my mind. It is likely that I will cease to be remembered when my own grandchildren have gone, though there is a distinct possibility that I may be lucky enough to survive in corporeal form long enough to see great-grandchildren grow to an age where I will be established in their memories. Now that is scary. How did I get to be so old so quickly?

In legal terms, a body is unlikely to be of forensic interest if the individual died more than seventy years ago. At the present time, seventy years takes us back into the middle of the Second World War. It is sobering to think that my great-grandparents, none of whom I ever met, are now technically archaeological skeletal samples and that my grandmother will be archaeological in less than thirty years from now—quite possibly within my own lifetime. Would I feel a sense of violation if someone chose to dig up my grandmother or great-grandmother to study them as archaeological specimens? You bet I would.

I wouldn't be too happy, either, about anyone poking about with the remains of my great-great-grandmother. Although the ties with our more distant ancestors are less strong and less visceral, in the minds of most of us, a blood bond still exists. So responsibility for treating archaeological remains with decency and dignity, and observing the sanctity of the need to leave a body at peace, must extend beyond the memories of our own lifespans. These are not just heaps of old bones, they are somebody's relatives, people who once laughed, loved and lived.

Recently I ran a workshop for some young people at Inverness College, in which we decided to take a closer look at a teaching skeleton that was hanging in their science laboratory. By the end of the day, once they knew they were actually face to face with a young man, no older than most of them, who was 5ft 4ins tall, had anaemia through poor diet and had probably come from India, they saw this skeleton in an entirely different light. They were no longer happy about him being put back in a cupboard and wanted him to be treated with greater respect. The anonymity of human remains deadens our empathetic responses but such is the power of forensic anthropology that it can reinstate identity and rekindle the human instinct to care and protect. I had hoped this was how they would react and they did not let me down. They were an incredibly mature and responsible group of young people.

Some remains will transcend any arbitrary definition of what is considered to be of forensic interest and what is archaeological, however long ago death may have occurred. There are important human considerations that make the barrier between those definitions porous—chiefly when the identity of a body that has been discovered is known or suspected, and the person's relatives are still living. For example, irrespective of the passage of time, no children's remains found on Saddleworth Moor, where the murderers Ian Brady and Myra Hindley buried their victims, will ever be viewed as anything other than of forensic relevance.

I was never destined to become an osteoarchaeologist, but that does not mean that I have not worked on archaeological skeletal material. I was first exposed to this in the fourth and final year of my undergraduate studies at Aberdeen University. After my third year—human body dissection, which I'd loved—I found myself faced with an odd mix of subjects that seemed to have been thrown together according to the interests of individual academics rather than to constitute any viable academic plan. I went from studying neuroanatomy one week to grappling with human evolution the next, then on to confocal microscopy (never understood that) and the unpleasant rhetorical musings of a rather sleazy academic who liked to talk about wetsuits and the douching effect they had on women. Bizarre.

Of more relevance was the obligation to undertake a research project in that final year. All the staff seemed to be researching in areas like lead levels in rat brain, carcinoma in the hamster pituitary or neuropathy in diabetic mice. I have a morbid fear of mice, rats and frankly anything rodentesque, dead or alive, so there was no way I was going to spend my research time with corpses of dead rodents. I pleaded and I begged with all of the academics to suggest almost anything else I might study. My future supervisor came up with the idea that I might like to consider identification from human bone for the purposes of forensic anthropology. Brilliant—no fur, tails or claws. No rapid scurrying movements, no biting, no scratching, and a natural progression from the human corpse in the dissecting room and the fresh meat of a butcher's shop.

I looked at how we might try to establish the sex of an

individual when presented only with fragments of a skeleton. The sample I was to use was a Bronze Age collection held in the museum at Marischal College. These archaeological remains were from the Beaker culture, named for their distinctive bell-shaped drinking vessels. It was their practice to bury these, sometimes along with some small stones or basic jewellery, in the cist (a stone-built box or ossuary) in which they interred their dead. In the north-east of Scotland these short cists were constructed with four upright stone side slabs and a horizontal capstone. Most had been unintentionally uncovered by farmers, usually when the edge of a plough lifted the capstone to reveal a skeleton crouched inside along with its beaker. It is thought that these people migrated from the Rhine area as traders and settled along the east coast of northern Britain. Because they were often buried in sand, the preservation of the remains was excellent and they formed a marvellous study collection for my research project.

The silent back rooms of the Marischal museum were a haven for me. Dusty, warm and smelling of wood and resin, they reminded me of my father's carpentry workshop. They gave me many quiet hours, hidden away between the archive stacks, to ponder the lives of the Beaker people, their health and how they died. These were peaceful folk and few of the deaths were traumatic. While I found them interesting, and was fascinated by the stories written in their bones, I was aware of a sense of incompleteness and a lack of fulfilment. This had less to do with the remoteness of a culture that existed 4,000 years ago than it did with the infuriating certainty that we would never really know the truth about their lives and deaths. It was all supposition and theory rather than fact. I was to find the lives and deaths of the more recent denizens of these islands more challenging, but more rewarding, too, once I brought the skills I'd learned to identifying the dead of today's world and answering some of the questions they pose.

As our islands have been inhabited for more than 12,000 years, it is inevitable that the working life of every forensic anthropologist will be crossed by archaeological material on a fairly frequent basis. Given the huge variation in population size over the centuries, we can only guess at how many people in total have expired on our soil but globally, it is believed that over a hundred billion people have lived and died since the appearance of Homo Sapiens about 50,000 years ago—fifteen times as many as the 7 billion or so of us alive in the world today. The living will never outnumber the dead because that would mean the global population expanding to somewhere in excess of 150 billion, which would not be sustainable.

In the twenty-first century one in every 39,000 head of population will die every day in the UK—that's over half a million bodies a year that must be 'managed', generally either by burial or cremation. There are only so many things you can do with dead bodies before they quickly become unpleasant to live with. Five traditional and accepted ways of dealing with them have been used by humankind around the world across the ages. First, they can be left exposed in the open for terrestrial and airborne scavengers to remove, the method still employed in the sky burials of Tibet. Secondly, they can be deposited in rivers or into the sea, where aquatic life will fulfil the same purpose. Thirdly, we may store our dead above ground, via immurement in mausoleums and the like, which has often been the preferred option of the wealthy. The fourth solution is to bury them in the ground, where the invertebrates of the soil will take on the scavenging processes. With the appropriate permission, we can technically bury a body anywhere we like, including on private land, as long as there is no risk of contamination of water sources. Lastly, we can burn them, which is currently seen as the quickest and most hygienic choice, though it does raise concerns about air pollution.

Perhaps the most extreme solution—and one neither advocated nor deemed socially acceptable today—would be to eat the deceased. While cannibalism (anthropophagy) is a feature of many cultures, in the UK evidence of the dead being used as a food source is sparse. Gough's Cave in Somerset, home at the end of the Ice Age to the Horse Hunters of Cheddar Gorge, is an exception. Skeletal remains found here show cuts consistent with the removal of flesh for consumption. There is more evidence in later centuries of medical cannibalism, arising from a belief among apothecaries in the mystical properties of the corpse. Preparations for ailments such as migraine, consumption and epilepsy, as well as general tonics, were made from various human parts. The rationale was that if death came upon somebody suddenly, their spirit could remain trapped within their corpse for long enough to bring vital benefits to those who chose to consume it. These 'corpse medicines' were often derived from ground-down bones, dried blood and rendered fat, along with many other equally unpalatable parts of the body.

A Franciscan apothecary from 1679 even gives us a recipe for human blood jam. He would first recover the fresh blood from recently dead people who exhibited a 'warm, moist temperament' and were preferably of 'plump build'. The blood was left to congeal into a 'dry, sticky mass' before being placed on a soft-wood table and cut into thin slices, allowing any liquids to drip away. It was then stirred into a batter on the stove and dried. While still warm, it was ground in a bronze mortar to a powder, which would be forced through fine silk. Once sealed in a jar, it could be reconstituted every spring with fresh, clear water and administered as a tonic.

Interestingly, according to one British academic lawyer, cannibalism is not in itself actually illegal in the UK, though mercifully there are laws against murder and desecrating a corpse. This revelation led my youngest daughter Anna, a trainee lawyer (or baby shark, as we call her), to muse, while sucking the blood from a cut to her finger, on whether any crime would be committed if one chose to eat oneself—a practice called autosarcophagy. And would consensual cannibalism be an offence provided nobody died? It seems that in UK law cannibalism is associated with the crime of murder, or at least of corpse desecration, rather than treated as a separate act. I worry about what kind of law Anna might go into.

Historically, in the UK ground burial has tended to be our favoured means of disposal. Ancient burial sites are likely to have been selected for the cultural importance or sacred significance of the land. As formalised religion took hold, burials were moved into churchyards or, if the departed was a sufficiently prominent figure, sometimes within churches themselves or directly below them in crypts.

With the mass migration to cities brought by the Industrial Revolution, we began to run out of burial space and the Victorian era saw the construction of municipal cemeteries, often in the outskirts of urban areas. Until the Burial Act of 1857, the reuse of graves had been common but as the cemeteries began to fill up, the eviction of some of their tenants rather more swiftly than was felt to be decent often led to public outrage. The legislation made it illegal for a grave to be disturbed, except when official exhumations were ordered. Interestingly, it was only opening a grave that constituted an offence. It was not against the law to actually steal a dead body—as long as it was naked.

Since the 1970s local councils have had the power to reuse longstanding graves provided that the coffin of the original inhabitant is kept intact. They have done so by deepening graves to make room for another burial on top. They would normally confine this practice to graves over a hundred years old which were untended, suggesting that they were no longer visited. In 2007, a shake-up of the Local Authorities Act in London, where the space problem is most acute, paved the way for boroughs to exhume remains and place them in smaller containers before reburying them, as long as the grave is at least seventy-five years old and there are no objections from leaseholders or relatives. This allows them to reclaim graves where there is room for more bodies not necessarily related to the original incumbent. In 2016 the Scottish Parliament enacted similar legislation.

Grave reuse remains an emotive issue and raises religious, cultural and ethical concerns. But with the shortage of burial spaces in the UK reaching crisis point—according to a 2013 BBC survey, half of all cemeteries in England will be full by 2033—something must be done to prevent the closure of cemeteries to any new occupants or we need to find another way to dispose of our dead.

With an estimated 55 million people dying every year worldwide, the problem is of, course, not restricted to the UK. The cities most affected are those that do not have a tradition of grave recycling. Durban, South Africa and Sydney, Australia, for example, have, like London, encountered strong cultural resistance to plans to introduce new legislation.

Many cities around the world, especially in Europe, have historically taken a slightly different approach, routinely removing bones from the ground or vaults and transferring them to vast underground catacombs or ossuaries, where the artistic skills of the custodian were given free rein. The largest of these is under the streets of Paris, where nearly 6 million skeletons lie, and perhaps one of the most ornate is the Sedlec ossuary in the Czech Republic, built in 1400 to house the skeletons removed from the church's overcrowded cemetery. In 1870 a wood-carver named Frantisek Rint was given the job of sorting out the accumulated heaps and began to transform the bones of between 40,000 and 70,000 people into outrageously elaborate decorations and furnishings for the chapel. There are chandeliers, coats of arms and fancy buttressing, all constructed from human bones. It seems that in his dedication to his art Rint allowed no sentiment to influence his choice of materials and viewing his handiwork can be an uncomfortable experience when you see how many of the bones come from very young children—including those frivolously used to create his signature.

In much of modern Europe the tradition of removing remains from cemeteries has naturally evolved into grave recycling. Germany and Belgium, for instance, provide public graves free of charge for around twenty years. After that, if families do not choose to pay to retain them, the occupants will be moved deeper into the ground or to another site, sometimes a mass grave. It is common practice in warmer climates, for example in Spain or Portugal, where bodies decompose more rapidly, for remains to be interred in the ground for a shorter period. If families then wish, the bones can be transferred to cemetery wall vaults for as long as payment is made. Ultimately, when there is no close family left, they are evicted. Some end up in museums where they can be studied and others are burned and ground down into ash. Singapore has a similar system to those used in Europe and Australia is about to adopt the UK's 'lift and deepen' option.

But burial, of whatever duration and whether in the ground or within monuments, is falling out of favour. The 30 million feet of wood, 1.6 million tons of concrete, 750,000 gallons of embalming fluid and 90,000 tons of steel that are buried underground in the United States alone are a stark illustration of its polluting effects. If those committed to the preservation of the planet are concerned about underground contamination by interments, they are no happier about cremation. Every cremation uses the equivalent of about 16 gallons of fuel and increases the global emission of mercury, dioxins and furans (a toxic compound). A broad estimate suggests that if you accumulated the amount of energy expended on cremations in one year in the USA alone, you could fuel a rocket for eighty-three return trips to the moon. Yet cremation is on the rise in the States—in 1960 it was the method chosen for only 3.5 per cent of deaths; today the figure is nearer 50 per cent.

Not surprisingly, the highest percentage is found in countries where cremation is the cultural norm or traditional choice for religious reasons, chiefly those with large Hindu or Buddhist populations. Japan tops the world cremation league table with 99.97 per cent, closely followed by Nepal (90 per cent) and India (85 per cent). In numerical terms, China has the most cremations—nearly 4.5 million a year.

Cremation burns away the organic components of the body leaving only pieces composed of dry and inert minerals, predominantly the calcium phosphates of the bones. The resulting ash represents approximately 3.5 per cent of the body and will weigh on average about 4lbs. In most crematoria, the remains are removed from the furnace and put through a mechanism called a cremulator, which grinds the bone remnants into ash and sifts out any foreign pieces of metal. In traditional Japanese cremations, the family pick the bone fragments out of the ashes with chopsticks and transfer them into an urn, starting at the feet end and finishing at the head so that the deceased is never upside down.

In the UK, about three quarters of the population now choose cremation over burial, but the rapid increase seen since the 1960s has levelled off in the last decade. Modern society likes to keep pushing the boundaries and newer 'greener' options are starting to emerge (cremated ashes are pretty much devoid of major nutrients). One is 'resomation', which involves alkaline hydrolysis. The body is placed into a vat with water and lye (caustic soda or sodium hydroxide) and heated to 160°C under high pressure for about three hours. This breaks down the body tissues into a greenish-brown liquid, rich in amino acids, peptides and salts. The remaining brittle bones are reduced to powder (principally calcium hydroxyapatite) by a cremulator and can then be scattered or used as fertiliser.

Another method, 'promession', works by freeze-drying the body in liquid nitrogen at -196°C and then vibrating it vigorously to explode it into particles. These are then dried in a chamber and any metal remnants are separated out with a magnet before the powder is interred in the top layers of soil, where bacteria will finish off the process. The latest green alternative, 'human composting', is still in the design phase but the idea is that a family will bring the body of their deceased loved one, wrapped in linen, to a 'recomposition' centre with a three-storey tower at its core—a giant version of a garden composter. Here the body is laid on woodchips and sawdust to aid decomposition. After four to six weeks, the body breaks down into about a cubic yard of compost, which can then be used to nurture trees and shrubs. They haven't yet figured out what to do with the bones and teeth, so perhaps human composting still has some way to go.

If such modern methods become the norm, fewer of us will leave behind as many traces of our physical selves as our ancestors did. Skeletal and other remains have enriched human history by giving archaeologists and anthropologists the voyeuristic luxury and academic stimulation of being able to study the people of former cultures at an up-close and very personal level.

While historical remains have usually consisted mainly of bones and accoutrements with which the dead have been buried, as we have discussed, certain climactic conditions—hot, dry heat, sub-zero temperatures or submersion, for example—have famously conserved some bodies almost in their entirety for centuries. Otzi the Iceman, discovered in 1991 in mountains on the border between Austria and Italy over 5,000 years after his death, had been nearly fully preserved, as was the body of John Torrington, from the ill-fated Franklin expedition of 1845, who was found 129 years later buried in the frozen tundra of northernmost Canada along with two of his colleagues.

'Bog bodies' such as Grauballe Man, Tollund Man, Lindow Man, Stidsholt Woman and the Kayhausen Boy owe the longevity of their remains to being buried in peat. Submersion in a mildly acidic liquid with high levels of magnesium was responsible for the remarkable preservation of the 2,000-year-old Han dynasty Chinese mummy known as the Lady of Dai, who was discovered in 1971 by workers digging an air-raid shelter for a hospital near Changsa. Even her blood vessels were intact, and found to contain a small amount of type A blood.

◊

Although it is rare for my team to have any significant involvement in the realm of archaeology, I was, against my better judgement, once persuaded, along with three fellow scientists, to participate in a BBC2 television series called *History Cold Case*, first broadcast in 2010–11. The formula was that we would be given archaeological human remains to examine, from which we were asked to piece together the lives they had once represented, with the researchers drip-feeding us information where appropriate. We genuinely hadn't a clue what we were going to be shown or what we would discover. So, while it was a little nerve-wracking, it was also quite intriguing. That said, I still moan and groan about it incessantly, mainly because performing in front of television cameras is not my thing—I have a face for radio. But all the stories we covered carried a reminder of just how far the dead, even from a very long time ago, can reach beyond their graves to touch us today.

The exposure and loss of anonymity that comes with appearing on television is a mixed blessing. It is unnerving to be considered public property by total strangers, whether they are approaching you to praise or criticise. Most people just want to tell you how much they enjoyed the programme but there are those who are not shy to comment on your looks, or something you said with which they vehemently disagree, or, of course, to point out that you are not very bright.

Three of the four presenters were women—something that was also remarked upon, and perhaps prompted more letters and emails of a personal nature than would be usual for the male of the species. Of the 'three witches of Dundee', as we were so aptly named, Xanthe Mallett, a forensic anthropologist and criminologist, was the target of quite a lot of inappropriate communications, but then, she is a very striking young lady. Caroline Wilkinson, our facial reconstruction specialist, would be sent gentle poems about the faces she restored and complimented on her skills as a sympathetic artist. As for me, I seemed to receive a disproportionate number of letters from inmates of HM prisons asking if I could help get them out, because 'honestly, it wasn't me who murdered my wife'. Our following within certain sectors of society also led to the programme being dubbed *Lesbian Cold Case*, which rather excluded poor old Wolfram Meier-Augenstein, our expert in isotope analysis, though I suspect the professor was not sorry to be overlooked.

On the plus side, we had many more lovely emails and letters from viewers who simply enjoyed discovering new things, and the contact with the public served to remind us that people are genuinely interested in learning about what the bodies of our ancestors can tell us and how we can use the science designed for the courtroom to help us delve into the lives of the past. There were many sad and poignant moments when we really felt that bringing back the stories of ordinary people, not the kings, the bishops or the warriors, but the children and the working girls, demonstrated that they had not been forgotten. Their stories had just been written in a language that required interpretation by forensic anthropology.

One sad case was the preserved anatomical specimen of a little boy of around eight years of age. This undocumented mummified child had been found in a cupboard in my department at Dundee University. His soft tissue had been dissected away, leaving only his skeleton and his artificially perfused arterial system. We knew nothing about him and didn't know what to do with him, so we hoped that the research carried out for the programme would lead us somewhere interesting.

It all started well but quickly became very dark. The child was not malnourished, and his death was not obviously medically explainable. The dating of his remains told us that he had died before the passing of the 1832 Anatomy Act. Could we be looking at the victim of one of the infamous child murders perpetrated at a time when anatomists would pay for a child's body by the inch? Or had he been stolen from his grave by the resurrectionists, the body-snatchers employed by anatomists to meet the demand for cadavers to train students and serve the interests of pioneering researchers? We know that the eminent anatomist William Hunter and the anatomist John Barclay were both performing vascular perfusions at that time and analysis of the chemicals present in the remains of our boy revealed them to be entirely consistent with those used by Hunter and his followers. The irony of anatomists uncovering the possible misdemeanours of previous anatomists was not lost on us.

At the end of the programme we were left with a decision we had not anticipated at the outset. What should we now do with this little boy's body? Should he remain within our department, go to a surgeon's museum or be given a proper burial? We were unanimous in opting for the latter. I have an aversion to seeing human remains being displayed like curiosities in a shop window for the titillation of onlookers. There is a fine line between education and entertainment and, in our hearts, we know when something is right and when it is wrong. The obvious yardstick is to imagine this was your son. What would you want? Unfortunately, it proved difficult to gain the necessary permissions to have him buried and he currently resides in a surgeons' museum, away from the public gaze, until his fate can finally be settled.

Another tragic figure was 'Crossbones Girl', a young woman in her late teens, almost certainly a prostitute, found in a pauper's grave in Cross Bones Cemetery in Southwark, south London. She had died horribly disfigured by tertiary syphilis, doubtless contracted through her occupation. Given the progress of the disease, we suspected that she couldn't have been much more than ten or twelve when she was first infected, which provided a chilling insight into the world of nineteenth-century child prostitution. When we reconstructed her face, the devastation wreaked by this condition on such a young person was shocking to see. Caroline then did a second reconstruction, showing how she would have looked if she had been healthy or could have been cured by penicillin. It is inevitable that we view an anonymous archaeological human skeleton with a certain amount of detachment but seeing the face of the young, flesh-and-blood woman more or less as she was, and as she could have been if fate had dealt her a better hand, dramatically brought home to everyone that we were dealing with a real person who had her own hopes, dreams and character; with a life that could be reconstructed almost to the point where we might have been able to return her name to her. Almost, but not quite.

The story that generated the biggest postbag concerned the skeletons of a woman and three babies excavated in Baldock, Hertfordshire, which dated from Roman times. The young woman had been found face down in a grave with the first set of newborn remains lying near her right shoulder. Further digging exposed a second neonatal skeleton between her legs. The third baby was still inside her pelvic cavity. What happened to her still happens today in many parts of the world where medicine is ill-equipped to deal with cephalopelvic disproportion (the disharmony between maternal pelvic dimensions and the size of the baby's head). It can also occur when the baby fails to turn in the uterus and presents in a breech position. Intervention to assist a birth that cannot take place naturally is relatively easy and safe nowadays, in developed countries, at least. But not in Roman Baldock.

The first child would have been successfully delivered, although we can never know whether it was born dead, or born alive and died shortly afterwards. The second triplet, who caused the problem, probably remained stuck in the birth canal, either because it was in the breech position (which was consistent with the arrangement of the skeleton) or because it just wouldn't fit through. It is likely that the mother died trying to give birth to this second child and was buried alongside her first baby. As she and the second triplet began to decompose there would have been a build-up of gases in her body which, assisted by a decompression of the baby's skull, finally succeeded in expelling the infant long after both their deaths in what is known as a 'coffin birth'. The third child never left the womb and died there, its exit blocked by the sibling stuck in the birth canal. What a heartbreaking outcome to an event that should have been a happy one but which instead resulted in four deaths.

More recently, my team at Dundee University assisted with a fascinating archaeological case in Ross-shire after a human skeleton turned up during the excavation of a cave at Rosemarkie, on the Black Isle just north of Inverness, a place suffused for me with memories of family days out in my childhood—especially that picnic when Uncle Willie got stuck in his beach chair. I like coincidences, and I really like it when times or places you have known in the past re-emerge in your life.

We agreed to undertake a study of the remains, using our forensic understanding of trauma analysis to uncover what had befallen this chap, for the Rosemarkie Caves project, a partnership set up between the North of Scotland Archaeology Society and the local community to investigate the archaeology of the caves, who used them, why and when.

The skeleton had been found under the sand at the back of what is known locally as Smelter's Cave, to the north of the Highland village. Radiocarbon-dating revealed that in all likelihood the man had lived during the Pictish period, before the arrival of the Vikings, in the late iron and early mediaeval ages. He was lying on his back in a 'butterfly' pose. His hips were flexed and his ankles crossed, splaying his knees. Between the knees a very large stone had been placed. His hands were on his waist or hips and stones had been placed along his arms. Another stone had been laid on his chest. The theory was that perhaps these were intended to keep the body weighed down to prevent it rising up in anger or retaliation, or perhaps just to ensure that he didn't float away on the tide.

Judging by the extent of the trauma to his skull, it was clear that he had met a violent end. There were no injuries to the rest of his body, and in all other respects he was a healthy, fit young man, probably in his thirties.

Trauma analysis is a logical deductive process that requires an appreciation of how bone behaves, how that behaviour alters when the bone is disrupted and subsequently suffers additional traumatic incidents, and how these can be sequenced. A possible implement, or implements, can then be identified. By looking at the position of the fractures and their relationship to each other, we can come up with a likely timeline of events showing the order in which the injuries were caused and what was used to inflict them.

It seems that the first assault on Rosemarkie Man was made to the right side of his mouth, where his teeth were smashed at the front as a result of being struck hard by some projectile, perhaps a spear or a lance or pole of some kind: it made a relatively neat and tidy entry and didn't penetrate all the way through to his vertebral column or appear to cause any further damage. He had definitely been alive when this occurred as the crown of one of his teeth was found in his chest cavity—in all likelihood he inhaled it after the impact.

Next came a very powerful wallop to the left side of his jaw, maybe from a fist, or perhaps the leading edge of a fighting stick, which would also fit with the circular shape of the fractures to the teeth on the right. This caused fractures to the main body of the jaw and at both joints where it articulates with the skull. The fracturing continued internally into the sphenoid bone at the base of the skull. The force of the second blow probably knocked the man backwards and, as he fell, his head made contact with a hard surface—perhaps the stones on the beach where he was buried. This set in motion multiple fractures that spread across his skull from the point of impact, which was slightly to the left-hand side of the back of his head.

While he lay on his right side his assailant, or assailants, evidently bent on making sure he didn't get up again, drove a rounded weapon, similar in size and shape to the one that fractured his teeth, into his skull through the temple behind his left eye. It exited in the same position on the opposite side, behind his right eye. The coup de grâce was a large, penetrating wound to the top of his head, delivered with such violence that it shattered the remaining parts of the cranium.

I was invited to Cromarty to present our findings to the local historical society. The skeleton had been found on the very last day of the excavations and the team decided to keep it under wraps to give the community an exciting surprise. Being from Inverness, I am quite well known in that part of the world, so there was a buzz of speculation as to why I was going to be attending their meeting. When, in the final slide of his presentation, the team leader unveiled a photograph of Rosemarkie Man in situ, an audible gasp went around the room. I then stood up, talked the audience through who our man was and what had happened to him and, finally, revealed a beautiful facial reconstruction created by my colleague Chris Rynn. The audience were thrilled.

Afterwards, one of the ladies told me she was so exhausted she was going home to have a lie down. Instead of the dry lecture on archaeological finds she had been expecting, she had been taken on a rollercoaster ride of emotions by the story of the brutal murder of a local man. She had even looked into the eyes of the victim, and at a face so lifelike that, though he had been dead for 1,400 years, it would not have been out of place on the streets of Rosemarkie that day. I just love the fact that humans cannot fail to be affected by the stories of other humans, even those who lived centuries ago, and how they embrace these forerunners as part of their neighbourhood because they once occupied the same patch of earth on our planet. People from Rosemarkie and the surrounding area even started sending us photos of their sons and grandsons, pointing out their resemblance to our Pictish man and suggesting that they might be related to him.

Such ancient archaeological studies bring huge satisfaction in terms of unravelling the complexities of the presentation of a body, but from the perspective of a forensic anthropologist they are frustrating, too, in that, no matter how certain we may be in our minds about how a person came to grief, there is nobody who can confirm whether we are right or give us any guidance as to where we might have gone wrong. As I first discovered as a young student while working on my Beaker culture project, it is the lack of evidential proof that is most vexing. For me, then, the more recent any foray into the archaeological world, the more rewarding it is likely to be as there is a greater chance of finding some documentary evidence that can help us to piece together more accurately the lives we are investigating and rebuild them on more solid foundations.

That is probably why I became so preoccupied with a quirky little nineteenth-century Irishman I encountered in 1991, when we excavated the crypt of St Barnabas Church in west Kensington, London. The ceiling of the vault was starting to crack and there was a genuine fear of collapse if the problem was not addressed. We were involved because the crypt had been used for burials and the bodies would have to be removed before the builders could come in to shore up the walls. We were given permission by the Archdiocese of London to carry out a recovery mission that would allow the coffins to be emptied, the bodies cremated and the ashes returned to consecrated ground.

The burials were in triple coffins typical of the early 1800s for those who could afford them. These multi-layered receptacles were like Russian Matryoshka dolls. There would be an outer wooden coffin, sometimes covered with fabric and featuring ornamental handles and other fittings in addition to a plate bearing the name and date of death of the occupant. Inside this was a lead coffin shell, sealed by a plumber and decorated with his bespoke patterning, also carrying a nameplate giving the details of the deceased. These lead coffins were designed to retain body fluids and were often lined with bran to soak up the foul decomposition liquor. They also ensured that the smell was contained and did not escape and waft upward into the church to offend the delicate noses of parishioners during their Sunday devotions.

Finally, there was a more perfunctory inner coffin made of cheaper wood, often elm, which functioned simply as a lining for the lead casket. Within this last box the deceased would lie in full repose, head resting on a pillow stuffed with horsehair, surrounded by cotton punched with holes to resemble more expensive broderie Anglaise fabric and usually dressed in their finest clothes.

By the time we arrived to excavate the remains, the outer coffins had all but disintegrated, leaving only remnants of wood and coffin furniture. The durable sealed lead coffins were another matter. We had to open these hugely heavy containers, like giant preserving tin cans, to get at the inner wooden coffins and remove what was left of the deceased. We had been granted permission to study and photograph the remains as a record of who had been interred there. The purpose of our research was to determine whether DNA could be extracted from these nineteenth-century burials. Could the genetic code survive lead coffin interment?

The answer, unfortunately, was no. As the body decomposes the fluids are slightly acidic. Because they had nowhere to drain, they had reacted with the wood of the inner coffin to form a weak humic acid, which strips the bonds between the base pairs (the building blocks of the DNA double helix) and the helical backbone. So the genomic information had dissolved into a thick, black soupy deposit at the bottom of the coffins, resembling a rich chocolate mousse (anatomists are prone to using food analogies to describe substances they encounter—not terribly appropriate, perhaps, but effective).

Given the number of early nineteenth-century burials and the proximity of the church to Kensington Barracks, it was not surprising that many of the coffin plates suggested strong military connections. Thanks to the various wars going on in Europe at this time, records from the period are exceptionally comprehensive. We had invited staff from the National Army Museum in Chelsea to keep a watching brief on our activities and to advise us on any incumbents who might be of historical note.

One burial in particular was of great interest to them. It was not the interred lady, Everilda Chesney, herself who caused the excitement but her husband, General Francis Rawdon Chesney of the Royal Artillery, celebrated for many achievements but especially for his epic descent of the Euphrates river in a steamer—a journey that demonstrated the possibilities of a new, shorter route to India that could cut out the long, treacherous voyage round the Cape of Good Hope. We left Everilda's coffin until last, just in case we got behind schedule, in the hope that the interest in her would buy us some extra time if necessary. We had only ten working days to open, record and transfer the incumbents of over sixty lead coffins.

Sadly, Everilda had died and been laid to rest in the crypt very shortly after her marriage. When we opened the coffin we found that most of her skeleton was fragmented. What were still intact were her delicate hand bones, inside a pair of finely crafted silk gloves. One hand was considerably larger than the other, which led us to suspect that she had suffered some form of paralysis in earlier life. If Everilda herself was unremarkable, the other contents of her coffin were interesting. Her eccentric husband had buried her with the full military uniform he had worn on their wedding day, 30 April 1839. He had laid two pairs of trousers across her legs, draped his military jacket across her chest and placed his forage cap near her head and his boots near her feet. The uniform was passed into the skilled curatorial hands of the National Army Museum and Everilda was duly cremated along with the other occupants of the crypt. All their ashes were returned for burial in consecrated ground. As time went on, I found myself becoming increasingly intrigued by Everilda's strange little husband (he was only about 5ft 4ins and had to wear cork inserts in his shoes to reach the height required for admission to the military academy). I read books in which he featured and started to research his life. One day I found a family website bearing his surname. Taking a deep breath, I posted a request asking if anyone had any information on the whereabouts of the diaries I had discovered he was known to have written. In response, I had the most wonderful email from Dave, a direct descendant of General Chesney, who lives near Chicago. And so began an online friendship that has continued now for over fifteen years. As I unearthed more details about his family, he would relay each new nugget to his ailing father, who would wait eagerly for the next instalment. 'Have you heard from the woman in Scotland?' he would ask Dave. 'What has she found out now?'

That a man who had been dead for over a century could be the catalyst for an enduring friendship between two people who have still never met, and provide a third person with a new interest in his twilight years, is really quite miraculous. There is no doubt that some characters from the past have the strength of personality to reach out far beyond their graves to touch contemporary lives. Skeletons are more than dusty, dry old relics: they are the footnote to a life lived, sometimes retaining sufficient resonance to ensnare the imagination of the living.

In Iraq after the second Gulf War, with the story of General Chesney percolating in my mind, I found myself one day sitting on the banks of the Euphrates river being guarded by none other than a battalion of the Royal Artillery. Another of those wonderful coincidences I so relish. Out of the blue, I heard myself say to the senior officer: 'Does the Royal Artillery have a benevolent fund?' I have no idea where the question came from and nobody was more surprised than me when it dropped out of my mouth. The lovely young man replied that they certainly did. As he went on to tell me enthusiastically about all the good work of the benevolent fund, a clear voice in my head was urging me to continue my research and perhaps to write down the story of the man behind the history. Maybe one day I will, and the Royal Artillery will benefit from my endeavours. I think Francis would approve of that.

I have to admit that I have a bit of a crush on my little Irishman and that what began as an interest has developed into a minor obsession: I once made my family go on holiday to Ireland so that I could find his grave and, through binoculars from a distance, lay eyes upon the house he built with his own fair hands. Fortunately, I have a very understanding husband who accepts that there are three people in our marriage.

CHAPTER 7

Not forgotten

'De mortuis nil nisi bene dicendum'
Of the dead, speak only good
Chilon of Sparta, Greek sage (600BC)

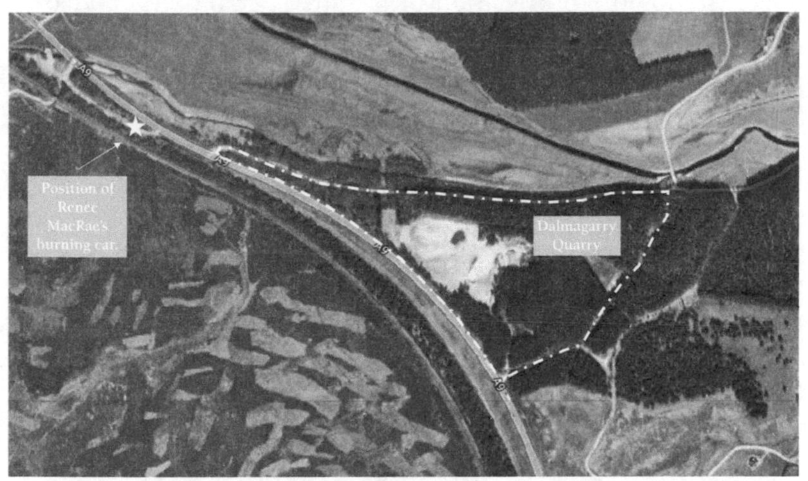

The position of Dalmagarry Quarry, the A9 road and the location of Renee MacRae's burning car.

Fascinating as archaeological remains may be, my heart lies in the here and now, trying to help solve more contemporary puzzles that will assist with the identification of the deceased or the prosecution of a guilty person who ended a life or damaged those of others. There is great satisfaction in finding answers for bereaved families and helping to bring perpetrators to justice, or confirming the innocence of someone wrongly accused.

As a student, having explored the world of the long dead and realising I was not inspired to linger there, I moved on, looking for an immediacy and an excitement that would challenge me at every turn and with every decision.

I never had any desire to work with the living. Although I appreciate the importance and the tremendous rewards to be gained from healing and tending to the sick, I always had a sneaking feeling that live patients would be more troublesome than dead ones. Being half control freak, half coward, I find that a more mono-directional interaction suits me best—in other words, a job where I am the only one asking the questions.

Had I chosen medicine, I know for certain that the first time I made a mistake that impacted negatively on the quality of someone's life or unnecessarily hastened their demise, I would have thrown in the towel. I would have lost all confidence in my decision-making ability and seen myself as a danger to my patients. Some would say that this is exactly the attitude all doctors should have, but I simply could not have continued if I thought I had done harm to someone else. So I think my road was mapped out from my teenage years. For me it was always the dead: all the way from the butcher's shop to the mortuary.

Forensic anthropologists don't always get it right, either, of course. That only happens in those dreadful CSI programmes where the clever-clogs scientist invariably triumphs in the end. What is carved most indelibly on our memories and on our perception of our reputations are those cases that remain unsolved, or where we feel we could have done more. And especially those where, no matter what lengths you have gone to, you cannot with certainty assign a name to an unidentified body, or where you have been unable to find a body but you know there is a strong likelihood that the missing person you are looking for is dead. Anything that prevents the circle being completed leaves you with a sense of unfinished business. These cases sit like mites under your skin and, however hard you scratch, you don't ever lose the itch until the mystery can be resolved.

I can think of nothing worse than not knowing where a loved one is or what has happened to them. Are they all right, or has some terrible misfortune befallen them? Are they dead, abandoned on some remote, lonely piece of scrubland or deliberately concealed in an anonymous hole in the ground? These are the thoughts that torture the parents, siblings, children, extended family and friends of the missing.

Grief is our response to any loss, not just a confirmed and accepted death, and those trapped in the limbo of not knowing whether someone dear to them is dead or alive often find it even harder to cope. It hits them every morning on waking, it is their last thought at night before they fall asleep, and on occasion it strays into their dreams as well. On the surface, with the passage of time, some learn to manage but, without warning, a name, a date, a photograph or a piece of music may at any moment catapult them back into the black pit of endless horrifying possibilities. This typifies the oscillation in the dual process model of grief between 'loss-oriented' and 'restoration-oriented' responses. A couple whose child disappeared once told me that it was as if their world had gone into a permanent stutter. While you are left replaying the same nightmare scenarios in your head in a continuous loop, you can never really begin to heal.

It is hard to imagine, too, the crippling, unresolved grief suffered by the bereaved who never have a body to mourn. Though they may be certain in their heads that the person they have lost is dead, their hearts may never acknowledge it. Families of those caught up in a fire, plane crash or natural disaster, for example, may have a reasonable expectation that a body will eventually be found, and having to accept that this is not always the case adds an extra burden to their grief.

This is why forensic anthropologists will examine every single fragment of a body, no matter how small, in an attempt to try to secure an identification. The case of a fatal fire in Scotland illustrates how we can make the vital difference between the fate of a person perhaps remaining for ever unconfirmed and a body being named and laid to rest. A remote house was destroyed by a blaze that had probably been raging for an hour or more before a farmer spotted the red glow in the distance and called the fire brigade. By the time the firefighters had been scrambled from a station over twenty miles away and had negotiated the winding, single-track country roads to reach the house, it was little more than a burned-out shell. The roof had collapsed, and the roof tiles and charred remnants of the contents of the attic space had buried everything under a layer of debris over 3ft deep.

The elderly lady who lived in the house was known to like a drink or two and was a heavy smoker. We were told that in the winter, to keep warm, she would usually sleep on a pull-out sofa bed in the sitting room, where she kept a coal fire burning day and night. We convened a forensic strategy meeting and agreed that her remains were therefore most likely to be found in the vicinity of the sofa bed. Once the fire brigade had certified the building safe to enter, we obtained a plan of the probable layout of the room and plotted the best way to get into it and to reach the sofa bed without disturbing vital evidence. Dressed in white 'Teletubby' suits, black wellingtons, face masks, knee pads and double-layer nitrile gloves, we inched our way painstakingly on hands and knees along the walls into the sitting room, clearing the rubble down to foundation floor level with brushes, shovels and buckets, searching all the while for the grey, telltale signs of fragmented bones.

It was slow work: the house was black from the fire, wet from the firefighters' hoses, still smoking in places and warm to the touch. After two hours we reached what was left of the sofa bed against the easterly wall and carefully cleared the debris on top of it, but there were no human remains to be found within its metal skeleton. We noted that it had not been unfolded, which indicated that it was unlikely the woman who lived here had retired for the night before the fire started.

After three hours, we had a second strategy meeting to decide where to search next. The vestiges of the sofa bed were removed and we debated whether we should continue further along the east wall or move west into the main body of the room. While surveying the destruction, I noticed a small, grey fragment no more than 3cm in length and about 2cm wide. We photographed and lifted it. It was part of a human mandible, with no teeth present, that had calcined or burned to such an extent that it was virtually ashed.

The suggestion now was that the remains might be lying between where the sofa had been and the fireplace. Here we recovered the highly friable and fragmented bones of a left leg, some vertebrae from the spine, melded with a nylon-type material, probably the remnants of the woman's clothing, and a left collarbone (clavicle).

So we had, it seemed, found the remains of the elderly occupant of the house. But how were we to confirm that they were hers? We were never going to be able to extract DNA from a skeleton that consisted of scarcely more than ashes. She had worn false teeth, but these had probably melted in the fire. We had to work with what we had. The clavicle held the key. This showed clear signs of having been fractured in the past. A bone that has been broken and has then healed rarely looks exactly the same as one that has never been broken. While a bone does mend, it is something of a patch job, and rarely does it repair itself with such precision that it leaves no clue to its previous misadventure.

The woman's medical records revealed that some ten years before, she had fallen and broken her left collarbone. It was enough for the procurator fiscal to allow her identity to be confirmed and grant permission for her remains to be released to her family for burial. There was barely enough to fill a small shoebox, but it was something.

For the fire recovery officers, this case was a wake-up call to the importance of having a forensic anthropologist at the scene. They admitted that they would never have recognised these grey lumps of ash as human remains; indeed, they might well never have noticed them at all, and just cleared them away with the rubble from the fire. Since that incident, in Scotland forensic anthropologists have regularly attended fatal fires along with the police and the fire service. A great working relationship has been forged and it has proven its worth time and again in the recovery of body parts that only a scientist could reasonably be expected to recognise.

◊

The two most troublesome categories of missing persons for our profession are those who go missing without leaving a clue as to where we might start to look for them and bodies to which we are unable to assign an identity.

We have all read newspaper articles about a young man or woman disappearing while walking home from a party late on a Saturday night. In this type of case, using research undertaken by the UK Missing Persons Bureau, we can speculate on what is most likely to have happened and activate the appropriate search protocols. For example, if a section of the route somebody took home was near water, perhaps a river, a canal or a lake, then these will be the places to check first. Approximately 600 people in the UK each year succumb to water-related deaths. The largest category (about 45 per cent) are accidental, around 30 per cent are suicides and less than 2 per cent are as a result of criminal intent. Perhaps not surprisingly, the day of the week most frequently associated with these deaths is Saturday, the peak time for recreational activities and excess drug or alcohol consumption. Some 30 per cent of deaths by water are coastal, shore or beach-related incidents, about 27 per cent are associated with rivers while the sea, harbours and canals account for roughly 8 per cent apiece. Of the reported suicides associated with water, over 85 per cent involve canals and rivers. Such compelling statistics explain why bodies of water are high on the list of prime potential search locations.

Research on missing children, too, provides invaluable information for major incident police teams and their expert advisers. Most children feared to have been kidnapped (over 80 per cent) are found swiftly and returned safely with no nefarious intent involved. Usually they have simply wandered off or got lost. Abduction murders, not surprisingly, receive extensive media coverage, but fortunately they are rare. The victims are more likely to be girls than boys, and few are children younger than five. These statistics provide no comfort to families who face this distressing situation, of course, but they are a necessary underpinning to pragmatic intelligence-led policing.

When a child is not found quickly, foul play does become the most likely explanation, although some families cling to the hope offered by tales of abducted children reunited years later, safe and well, with their parents. Such cases are unusual but they are not unheard of, as the story of Kamiyah Mobley illustrates. Taken from a hospital in Jacksonville, Florida, in 1998, when she was only a few hours old, by a woman who had recently suffered a miscarriage, Kamiyah was eventually found alive and well eighteen years later, 300 miles away in South Carolina, having enjoyed a generally happy childhood oblivious of her true identity. For only a very few lucky families will there be an outcome such as this, and it comes with a heavy price, not least of which is the foundation-shaking damage caused to the child's sense of identity and belonging. For other children, abducted with more sinister intent, there may be a legacy of abuse—every parent's greatest nightmare.

In spite of being well aware that stories like Kamiyah's are exceptional, many affected families keep a small flame of hope alive across many decades, which perhaps helps to numb the hard edges of their pain. Without a body, and especially when there is no evidence to confirm that their child is dead, they consider that to do otherwise would be tantamount to abandonment.

Such cases remain officially open, waiting patiently for some new evidence to surface, for as long as there is public benefit in their being so: a surviving family, or the chance that the perpetrator may still be alive to be brought to justice. As a police superintendent reminded me recently: 'There is no such thing as a closed cold case.' When a body is found and we can make a positive identification, the news is never welcome to relatives as it dashes those long-nurtured hopes and dreams and forces a harsh acceptance of the reality of the ultimate loss. And we are only too aware of the further pain that will be caused as the investigation reveals the circumstances surrounding the last days and the death of someone precious to them. But I like to think that, in the long term, uncovering the truth will prove to be a small kindness in finally breaking that stutter of uncertainty and permitting some level of coping and healing to enter their lives.

I think often about the families whose children are still missing and wonder how I would feel if I had to walk in their shoes. In general I have tried to maintain, as far as I can, anonymity for those whose personal tragedies are recounted in this book but one exception I would like to make is for two missing children, and one mother, who have never been found, in the hope that maybe, just maybe, revisiting their cases might help to find and return them to those who still miss them. Their families have accepted that they are dead, and their one desire now is to know where their loved ones are and to be able to 'bring them home'. Who knows what turn of events might suddenly jolt someone's memory or conscience, and if there is the tiniest chance that telling their stories again here could help bring two families the answers they so desperately need, then it is worthwhile. My grandmother, a great believer in fate, taught me that we never know when an alignment of moments might produce the right alchemy for change.

The first disappearance dates back to my teenage years, and I remember it vividly because it happened on my own doorstep. I could never have imagined then that nearly thirty years later I would become involved in what is currently one of the UK's longest-running missing persons cases. Renee MacRae, aged thirty-six, of Inverness, and her three-year-old son Andrew were last seen alive on Friday 12 November 1976. Police were initially told that after dropping off her elder son with her estranged husband, she was heading for Kilmarnock to visit her sister, though it later emerged that she was probably going to meet the man with whom she had been having a four-year affair—and who, it transpired, was Andrew's biological father—William MacDowell.

That night, twelve miles south of Inverness, a train driver noticed a car on fire in a layby on the A9. It was Renee's blue BMW. The car had been destroyed by the time the fire brigade arrived and there was no sign of either Renee or her son. No trace of them was found in the car, either, although a blood stain later identified in the boot was of the same blood group as Renee's. All sorts of wild rumours did the rounds at the time, including outlandish tales of planes landing at Dalcross Airport with their lights switched off and claims that Renee had been whisked off to a life of luxury in the Middle East, abducted by a rich Arab oil sheikh.

Of course, there was absolutely no basis to any of it. This is simply how, since time immemorial, communities have responded to such inexplicable and devastating events, spinning yarns and building myths around them that become part of local lore. Such stories will be reported by well-meaning but misguided citizens as well as by those just eager to grab a bit of the limelight. Whatever the intention, they rarely assist and often waste a great deal of valuable police time.

I remember the police coming to our door one Sunday afternoon and talking to my father. Over a hundred officers had been drafted in to search for Renee and Andrew, supported by several hundred local volunteers and army reservists. They were checking every rural outhouse, shed and shack and we were searched like everybody else who lived out in the countryside in the vicinity of the A9. There was not a household in Inverness that was not affected in some way or other by the disappearance of Renee and her son. The police were out day and night, combing Culloden Moor and all the buildings and scrubland nearby. The RAF flew Canberra bombers, equipped with heat-seeking devices, over the area and divers investigated lochs and flooded quarries. No stone was left unturned.

A detective sergeant working on the case, who had started some excavation at Dalmagarry Quarry just north of Tomatin, only a few hundred yards from where Renee's car was found, reported a stench of decay. But for some reason the digging was halted and soon, with no fresh lines of inquiry to follow, the police started to wind down the case. A huge incident like this leaves a scar on a community that never truly heals. It is impossible for whole towns to move on, let alone the immediate family and friends left behind, until the missing are found. When a young child is involved, the poignancy of the loss remains sharp across the decades. If Andrew were alive today he would be in his forties (Renee would be in her seventies), and every time a significant anniversary of their disappearance approaches, the local press will retell their story. At first glance this might seem a little macabre, but it serves to keep the case active in the public consciousness.

In 2004, the construction of a new stretch of dual carriageway on the A9, for which a supply of sand and gravel was needed from Dalmagarry Quarry, provided the police with an opportunity to reinvestigate the quarry and the surrounding area and to finally close off a part of the original inquiry over which questions remained.

Dalmagarry Quarry occupies an isolated triangle of land of about 900 square metres between the A9 to the south-west and the steep slope down to the Funtack Burn and the Ruthven road to the north. This location was the source of several pieces of circumstantial evidence from 1976. A member of the public had reported seeing someone walking on the A9 that evening in the dark, possibly with a pushchair (Andrew's was never found). There had also been sightings of a person hauling what looked like a dead sheep up the slope towards the quarry (Renee was said to have been wearing a sheepskin coat on the night she disappeared). By coincidence, the man with whom Renee was having an extra-marital affair, Bill MacDowell, worked for the company that had been cutting the quarry at the time. When these small pieces of information were added together, along with the report of the DS who had smelled decomposition there after starting the 1976 dig, there was enough to justify a new, full investigation of the site as part of a cold-case review.

I was asked, along with the country's leading forensic archaeologist, Professor John Hunter, to lead the excavation in an attempt to find any evidence of the remains of Renee and Andrew MacRae. Aerial images taken by the RAF in 1976 allowed us to determine the precise morphology of the different faces of the quarry at that time and to dig it back and reconstruct it precisely as it would have looked during that period of activity. Once this excavation had been achieved, we could look for areas where remains might have been buried. We worked in conjunction with the owners of the site, and they provided diggers and expert drivers who would become an integral part of our forensic team.

Dalmagarry Quarry is a desolate place, accessible only via a gated track off the A9 that was permanently guarded for us by the police. The media attention was fierce and some overzealous members of the public did what they could to influence our thinking. A few, convinced that the police were concealing information—which they most certainly were not—even tried to pressurise us in their attempts to find out what was going on. Thank goodness this was before the advent of drones. We held a press day to explain what we were hoping to achieve, promising to update the media as and when there were any developments, and crossed our fingers and toes that it would be enough to keep them happy and ensure that we would be left in peace to get on with our work. As it turned out, the dig went on for so long that I think they eventually got bored and forgot we were there.

The excavation inevitably prompted a barrage of what Viv calls 'File 13' correspondence—letters from both conspiracy theorists and well-intentioned members of the public motivated to involve themselves in a case in the mistaken belief that their pet theories and fantasies will provide the vital piece of evidence that eventually solves the mystery. I had letters telling me to dig in specific locations under the A9—one correspondent had even been out on the road and marked a yellow X on the tarmac to show me where. I was informed that there were local gangs of human traffickers and paedophile rings run by the police, which was why we would never find the bodies. Many named their prime suspect and urged us to go and dig up the horse paddock at his home. And of course I had mail by the bagful from clairvoyants. All I can say is that the spirit world must have been having fun with them because none of them offered the same answer. I know most of these people were only trying to help, but in practice such letters only ever take up time and generally produce nothing of any relevance or value.

In the thirty years since Renee and Andrew had gone missing, the quarry had been filled, levelled and planted with trees. We estimated that it would take us at least a month to dig it back to its 1970s profile and to highlight any areas of possible interest. If remains were found, then it was likely we would need longer.

The first job was to clear the site of around 2,000 trees to expose the current ground level and map out the superimposition of the quarry as it had been. The speed with which the trees were cut, stripped and sectioned by the modern forestry harvester equipment was simply incredible. A job that in the past would have taken weeks was finished in a couple of days. Some trees were left strategically placed along the sides of the quarry to provide some shelter and privacy and prevent covert media shots and curious members of the public rubber-necking from the already dangerous A9. I was confident that if Renee and Andrew were here, we would find them, though I immediately regretted my enthusiasm in saying so to the media. It was never my intention to raise false hopes. If we were not successful, at the very least the site could be written off as an area of interest in this case.

It was also possible that the quarry could have been used as a primary deposition site. The remains could have been hidden there first and later lifted and transferred to a secondary or even tertiary location. This theory was consistent with all the contemporary intelligence apart from the reported smell of decomposition. Primary deposition sites are generally chosen for their convenience and proximity to the crime scene (possibly, in this case, the burning car). They also tend to be familiar to the perpetrator. As most murders are not planned, there is often an element of initial panic in disposing of the body and ancillary evidence. Once the killer has had time to think, he or she may return to the primary site and move the remains to a safer place, usually further away from the crime scene. Because they have been better thought out, secondary deposition sites are much more difficult to predict or find and tertiary sites even harder.

The next four weeks were spent shifting over 20,000 tons of earth from the quarry, co-ordinating the work of the diggers with that of the forensic archaeologists and anthropologists, looking for bones, clothing, bits of pushchair, luggage and so on, bucketload by bucketload. The archaeologist directed the digger and inspected the surface of the soil after every scraping and the anthropologist searched through every bucketful. We had dry weather, wet weather, hot and cold weather, hailstones and biting winds—all in the same day sometimes.

What did we achieve? We knew that we had restored the

topography of the quarry to exactly how it had been in 1976 and earlier. We found items confirming that, including an empty packet of salt and vinegar crisps advertising a Queen's Jubilee competition promoted by Jimmy Savile. We knew that if remains had been buried there we would have found them, as our minute screening turned up much smaller bones than those we were looking for, such as those of rabbits and birds. We discovered the potential source of the rotting smell: the site where rubbish and waste from workmen's toilet facilities had been buried during the construction of the A9 in the 1970s.

But we did not find Renee MacRae, we did not find Andrew and we did not find any circumstantial evidence relating to either of them or to their disappearance. It was hugely deflating for the team that had gone into this mammoth operation with such high hopes, but we knew we had done our best and we were confident that, wherever they had been and wherever they were now, they were not in Dalmagarry Quarry.

The cost of the excavation is thought to have been over £110,000 and it would have been a small price to pay if we had found any trace of Renee and Andrew. The chief constable at the time took some quite extensive flak for his decision to investigate the quarry nearly thirty years after their disappearance, but he would have been a hero if their remains had been found. Personally, I think it was a brave and bold decision and one that demonstrated unswerving police commitment to closing such cases, regardless of the passage of time.

Back in my office, reflecting on the excavation and mulling over what else we could have done, I was deeply touched to receive a handwritten letter from Renee's sister, thanking us for our efforts. Her one wish is not retribution but to have her sister back, to be able to bury her with dignity and to know that she is home and safe at last: the universal desire expressed by all families unfortunate enough to find themselves condemned to spend their lives waiting for the knock on the door that might just be the harbinger of astonishing, elating news but which they know is much more likely to bring long-anticipated heartbreak.

When these searches are successful, it is, of course, exhilarating for us. When they do not produce what we are seeking, we simply have to accept that we are looking in the wrong place and cannot find what was never there. Renee's sister summed this up much more eloquently than I ever could in an interview: 'Time can never heal the pain, and I can't believe that time will ease the conscience so much that someone out there can believe they will get away with murder. It always gives me some hope when I read of an old crime being solved. Maybe one day.'

Time, patience and conscience are the ingredients that fuel hope for the families of the missing. Police Scotland have not given up on Renee and Andrew MacRae and neither has their family. Somebody out there knows what happened to them and where the bodies lie. Perhaps they have kept silent all these years about something they know or have heard about, reluctant to point the finger of blame. But as time passes, allegiances change, relatives and acquaintances die, and if this person, or people, have a conscience, even if they heed it only on their own deathbed, they must do the decent thing and put an end to the family's misery.

◊

The second case I want to highlight is that of eleven-year-old Moira Anderson, who popped out from her grandmother's house in Coatbridge on a cold winter's day in 1957 to buy butter and a birthday card for her mother and has never been seen since. In a break with the normal protocol, in 2014 the lord advocate, Frank Mulholland, named her killer as paedophile Alexander Gartshore, who had died in 2006, forty-nine years after Moira's disappearance. Bus driver Gartshore, the last person known to have seen Moira alive, was indicted for her murder, which is not, of course, the same thing as being found guilty. Technically, he remains innocent until proven guilty in a court of law, but as he is no longer alive to face trial, that can never happen.

I remember, in 2002, sitting with a retired murder squad officer watching news coverage on television of the investigation into the disappearance of schoolgirls Holly Wells and Jessica Chapman in Soham, Cambridgeshire. Their school caretaker, Ian Huntley, who was maintaining that he had spoken to them as they walked past his house, was being interviewed by a news crew. The former detective remarked to me: 'Always look closely at the person who claims to be the last to have seen the missing person alive. He looks dodgy to me.' As we all now know, it ultimately transpired that Huntley had murdered Holly and Jessica. I was in awe of this officer's prescience. Police instincts, allied to years of experience, can be priceless. So many elements of police investigations are driven by technology these days but good, old-fashioned detective work should never be allowed to go out of currency.

Key to the background of the Moira Anderson case is the most amazing campaigner called Sandra Brown. Sandra, who was a few years younger than Moira and grew up in Coatbridge at the same time, has doggedly pursued the truth of what happened that day. As well as campaigning tirelessly on child-protection issues, in 2000 she set up the Moira Anderson Foundation, which helps families affected by child sexual abuse, violence, bullying and related problems. In 1998 she wrote a book, *Where There is Evil*, about Moira's disappearance and the investigations of the previous forty years. An inspirational example of sheer determination to see justice served in the face of monumental adversity, it examines, in classic no-nonsense Lanarkshire style, but with compassion and empathy, the devastating effect of child abuse on everyone who comes into contact with this most heinous of crimes.

Sandra believes that there was a well-protected paedophile ring active in Coatbridge at that time and named Alexander Gartshore as the person responsible not only for Moira's abduction but also for her murder. What is so utterly astonishing about Sandra's campaign is that Alexander Gartshore was her father.

I first met Sandra in 2004, when Gartshore was still alive and she was actively pursuing any and every lead she could. She contacted me after bringing a psychic into the search for Moira (yep, the spooks are always there, aren't they?). They had found some bones. Would I look at them?

The bones had been discovered while they were out looking for Moira's remains near the Monkland Canal. The psychic had apparently been overwhelmed by the force of the pain and distress emanating from them. There was no doubt in his mind that the bones were channelling the pain and suffering of a child, who he strongly believed to be Moira.

My views on this sort of thing are fairly unambiguous: it's absolute nonsense. I think I do understand, though, why people bring in self-proclaimed psychics, especially when all else has failed and they feel they have nothing to lose. Some of these 'psychics' are earnest but misguided; others are charlatans, and I worry about the damage they can do to vulnerable loved ones. But since bones had been discovered, I agreed, making it very clear to Sandra that if these remains did turn out to be human then there could be no further contact between us as this would become a police matter. She respected that and understood fully. Sandra is a dear friend now and I know she will laugh at this, but I wondered if perhaps she was a bit doolally.

The delivery of the bones was arranged in a very cloak-and-dagger manner. The psychic apparently worked at the University of Dundee—now there's a coincidence—but, I was told, wanted to remain anonymous, so would leave them outside my office door. I waited for these bones to appear and one day they did. For relics holding so much power and pain they had been treated with scant respect, just shoved into a supermarket carrier bag and left hanging on my door handle. A note on the bag read simply: 'Monkland'. Before opening it, I made notes, took photos of the bag and donned mask and gloves to ensure that, if these were human remains, there would be no DNA cross-contamination. I admit to being a little nervous as I unwrapped them. But within seconds I was exclaiming under my breath, 'Oh, for the love of the wee man!' as I found myself looking at the butchered rib and shoulder bone of a large cow.

I broke the news to Sandra and she took it like the trouper she is. To her it was just another avenue closed and she would carry on with her mission. I was in touch with her sporadically over the next few years as she, the Moira Anderson Foundation and Moira's family maintained their assault on the legal and investigative authorities. Around 2007 she started to talk to me about the possibility of Moira's body being buried in a grave in the Old Monkland Cemetery. She had approached the lord advocate of Scotland, Frank Mulholland's predecessor, about examining the grave in question, and felt their discussions had been encouraging.

As negotiations continued through 2008 and 2009, Sandra arranged for Moira's sisters to have their DNA samples taken and analysed and asked me to store the reports in case they were needed. I still hold them to this day. She provided me with a comprehensive list of what Moira had been wearing when she disappeared so that, should we find the buttons from her coat, the buckles of her shoes or her Brownie badge, we would be aware of their significance. Now in full battle mode, she sought and was granted permission for a GPR (ground-penetrating radar) survey of the grave. I do not profess to understand GPR visuals, but the results certainly appeared to show some anomalies of interest. Having said that, this was a graveyard, which is exactly where you would expect to find holes being dug and human remains being buried.

In 2011, I had a long meeting with Sandra and she talked me through the reasons why she wanted us to excavate the grave and undertake some exhumations. She believed that Moira's body might be found directly beneath the coffin of a Mr Sinclair Upton, who had been buried there on 19 March 1957.

The theory was that Moira had met her death at the hands of the suspected paedophile ring on or around 23 February, the day she went missing, and that her body had perhaps been hidden somewhere, possibly within a well-hidden compartment in the bus Gartshore drove, until a suitable disposal site could be found. If he was involved, getting rid of her body would have been a matter of particular urgency for Gartshore, as he was shortly due to appear before the Coatbridge Sheriff Court to face charges relating to the abuse of a twelve-year-old and he would have been anticipating a prison sentence. Indeed, on 18 April, he was given eighteen months in Saughton Prison. It was while serving this sentence that he made a comment to a fellow inmate about how a late acquaintance of his, 'Sinky', had 'done him the greatest of favours that he never knew about'.

Sinclair Upton, who had died aged eighty the month before Gartshore's incarceration, was a distant relative, and Gartshore would have been aware of his death and impending burial at the Monkland cemetery. Had the death of this innocent man provided a timely and secure disposal site for Moira? Here was a hole in the ground, ready and waiting, and it would have been ideal: who would go looking for a missing body in a cemetery? Gartshore would have known that the lair would probably be dug and left open over the weekend ready for the funeral on the Tuesday. Could Moira have been placed in the lair, perhaps covered by a thin layer of soil, before Mr Upton's coffin was lowered down on top of her, thereby concealing her perhaps for ever?

I have to say that Sandra's research and her logic were compelling. I produced a plan for the Crown Office detailing the work that would need to be undertaken to examine the grave and then we sat and waited. That year Frank Mulholland QC, now Lord Mulholland, was appointed lord advocate. A big man who did not shirk controversial decisions, Frank was very hands-on and, having been born in Coatbridge himself, only two years after Moira went missing, he had a strong connection to the town and understood the community's need for resolution. In 2012, he ordered cold-case detectives to reopen the inquiry as a murder investigation, we got the go-ahead to talk with DCI Pat Campbell of what was then Strathclyde Police and began to discuss exhumations. We had now been on this case with Sandra for eight years.

The agreement was that, with the full permission of the local council and all the families involved, we would commence the exhumations, maintaining full focus on the potential requirement for a switch to a forensic investigation should we identify juvenile human remains, as there were none recorded in this burial plot. To this end, we would be supported by Strathclyde Police and, if the nature of the exhumation did change, the case would come under the direction of the Crown Office. We would then immediately cease to be working for the family and would instead be working for the Crown.

We advised that it would be best to schedule the exhumations for the summer, when the days are longer, there is less rain, the weather is warmer and the soil at Old Monkland Cemetery, which is heavy with clay, would be drier, making the digging easier. As it turned out, the paperwork was at last completed in December. So when were we asked to start? In the second week of January. My colleague Dr Lucina Hackman and I are thinking about advising the police in future to plan digs for the winter months, in the hope that reverse logic will achieve the desired result. Plain talking doesn't seem to work.

We established that there ought to be seven coffins in total in the triple-width family burial plot. Three on the left, which were more recent, interred in 1978, 1985 and 1995 respectively; one in the middle, dating back to 1923, and three on the right, where the records indicated Mr Upton's coffin should be lying between that of his wife, laid to rest in 1951, and a later burial in 1976. There was no justification at the outset for disturbing the remains on the left-hand side or in the centre of the plot, although we had permission to do so if this became necessary, for example, if we discovered that Mr Upton's coffin was not where we expected to find it.

Coffins do not always end up where they are supposed to be. Occasionally people get buried in the wrong place for a variety of reasons. Sometimes when a grave is opened it turns out that there is insufficient space to put the coffin where intended; sometimes they are put in the wrong place purely by mistake. Records do not always accurately reflect the true picture. Indeed, after my own grandmother died in 1976 and we opened the grave for her to be buried with her husband, we found a child's coffin in our lair. As far as any of us knew, no children in our family had died and been interred there, and when we checked with the cemetery there was no record of any additional burial. It happens. The child was moved to consecrated ground elsewhere. I did feel a little uncomfortable about it but my grandmother had to go somewhere, and space needed to be kept at the top for my father, when his time came.

I had a dream team for the job at Old Monkland Cemetery. Dr Lucina Hackman has been with me at Dundee for sixteen years, Dr Craig Cunningham for over ten and Dr Jan Bikker I have known since he was a PhD student. Having developed a high level of mutual trust and respect, we are so used to working together that we are all attuned to the tasks the others are performing and able to anticipate what our colleagues need without a word being spoken.

Before we began we had to ensure that the memorial headstone was secured; in the end, we had to remove it temporarily because of the risk of it keeling over into the hole on top of us and adding another four dead bodies to those already in the ground. To gain access to Mr Upton's coffin we first had to dig down to Mrs McNeilly, who had been buried in 1976. The clay-based soil was solid and although machinery was brought in to scrape off the surface layers down to the lid of her coffin, the rest had to be dug by hand in case this became a forensic investigation.

We had permission to undertake a brief anthropological examination to determine whether the occupant of the first coffin fitted with the description and age of Mrs McNeilly, who had been seventy-six when she died. The coffin was classic 1976 thin veneer with a chipboard shell and, given the degree of waterlogging of the soil, we knew it was likely to be in a poor state of repair. We found that indeed it was. Mrs McNeilly was carefully removed and placed in a heavy-duty body bag, which was housed securely until she could be laid in a new coffin and returned to her resting place. We were satisfied that her remains were consistent with her identity.

With barely six hours of daylight each day we needed generators and lights to complete a normal ten-hour shift. Heaters would have been brilliant but they never materialised. It was a bitter west of Scotland winter, freezing cold and so, so wet. As we worked in the sodden clay, we began to sink. If we tried to step back we'd find we'd left our wellies behind, sucked into the mire, so our feet were permanently muddy, wet and cold. Such conditions make for a very long, miserable day. Anyone who thinks forensic anthropology is sexy should spend a day in Old Monkland Cemetery in January, chilled to the bone, up to their knees in mud and clay with the excavation walls continually in danger of collapsing around them and creating their own tomb.

When we lifted the base plate of Mrs McNeilly's coffin we expected the next discovery to be the lid of Mr Upton's.

Sure enough, a glimpse of a metallic sheen and a change in the sound made by the spade as it hit wood indicated that we were there. The coffin was solid wood, as was usual for its time, and perfectly intact. The metal turned out to be the fragile remains of a coffin name-plate. This was carefully removed and dried slowly so that it could be cleaned, which revealed his name, age and the month and year of his death. Mr Upton was exactly where he should have been and all the information was correct. What we did not know was whether Moira might be within the coffin, underneath it, to the side of it or indeed possibly in the coffin below his, in which his wife had been interred six years before him. All were viable options, given the possible scenarios, and so all had to be investigated.

We removed the lid of Mr Upton's coffin to find his perfectly preserved skeleton. We laid out his remains meticulously to ensure that there were no juvenile bones present (there were not) and he, too, was then placed, bone by bone, in a heavy-duty body bag and stored safely until he could be returned to his lair. The sides of his coffin were delicately dismantled to expose the base plate. The most likely place for Moira to have been concealed now, if Sandra's theory was correct, was under the base plate of Mr Upton's coffin and above the lid of that of his wife. As we lifted the base plate, we found there was barely space for a cigarette paper to be slipped between the two. Wherever Moira was, she was neither in Mr Upton's coffin nor in the gap between his and Mrs Upton's.

This did not mean, though, that she had not been 'tucked' around the sides of the grave, so we excavated outward laterally and to the head and foot ends of the coffin space. Nothing. Our last task was to examine Mrs Upton's coffin. When the grave had been opened prior to Mr Upton's funeral, it was possible that his wife's coffin could have been broken into and Moira placed inside. Like her husband's, Mrs Upton's coffin was in perfect condition. When we lifted the lid, all we found were the remains of an elderly lady. We cleared around the margins of this coffin, too, but again discovered nothing. Wherever Moira is, she is not in that grave in Old Monkland Cemetery.

Having to break the news to Sandra, who had been so hopeful of a resolution for Moira's family and the whole Coatbridge community, and whose strong conviction had driven her to campaign so tirelessly for so long for the grave to be excavated, was hard. It was also distressing for Mr Upton's family, unwittingly caught up in a case with which they had little connection. Their involvement is an illustration of how far the ripples of such events spread out. They, too, had hoped Moira would be found in the grave as it would have made the exhumation worth the upset it caused. As it was, they shared the disappointment felt by everyone in their town. Their relatives were reburied and the family held a memorial service at the graveside.

The little girl who vanished into thin air in Coatbridge in 1957 remains missing and her case remains open. Given that she has been missing for sixty years, the number of people alive who might hold key information is diminishing. It is unlikely that we will be able to prosecute anyone else now in connection with her disappearance, but the race is on to provide some peace to her elderly sisters by finally bringing home their wee sister.

Very recently, the cold-case team drained and searched an area of the Monkland Canal after receiving information suggesting that on the night in question someone had been seen throwing a sack into the water. Radar showed some anomalies in the canal floor and divers were sent in to investigate. Our team was again present, but all we had to identify were the bones of a large dog, possibly an Alsatian. Other locations will be considered and maybe one day we will get lucky, through either intelligence or happenstance.

In the case of Renee and Andrew MacRae, who disappeared

twenty years later than Moira, there is a little more time for someone to come forward and ease the suffering of their family. The not knowing is one of the most debilitating burdens for those grieving for the missing. If the work we do brings them some little comfort and relief, then it has great value. And if, by chance, the perpetrators of crimes that become cold cases such as these are still alive, they can be brought to justice. There is no statute of limitations on the crime of murder.

CHAPTER 8

Invenerunt corpus—body found!

'True identity theft is not financial.
It's not in cyberspace. It's spiritual'

Stephen Covey, educator (1932–2012)

A facial reconstruction of the man from Balmore.

Without a body, investigating what has happened to the missing can be extremely difficult. It can be just as problematic when a body is found and there are no obvious clues as to the person's identity.

Unfortunately, as we have seen, the idea that for every unidentified body there will always be a corresponding report of a missing person, and all we need do is connect the two, is a vast over-simplification of reality. A report may have been made in a different country, to a police force distant from where the body was found, or may have been recorded many years before, archived and forgotten. Perhaps there has been no report at all because nobody realised the person was missing, or there was no one who cared sufficiently to raise the alarm. Some may see that as a sad indictment of society but the fact is that some people do not want to interact with others or to be part of a community, and as long as they are doing nothing illegal, their right to privacy and anonymity has to be respected. When those who prefer to live alone and unknown also die alone and unknown, reconciling them with their identity can be challenging, and, in some cases, sadly, unachievable.

A time lapse between death and the body being found can complicate matters. We were once called to a council flat in London occupied by a Chinese gentleman. He had not paid his rent in over eighteen months and eventually the council entered the property to repossess it. They were shocked to find the tenant in bed, wrapped tightly in his duvet like a cocoon. He had died in his sleep over a year before and was almost totally skeletonised. The bedding and mattress had acted like a wick, drawing away all the moisture produced by decomposition, leaving the remaining soft tissue desiccated and effectively mummifying him.

This man had lived alone and died alone, unmissed and anonymous. Neighbours told the police conducting door-to-door inquiries that they hadn't registered his absence, though some said that, come to think of it, they had noticed rather a lot of dead flies on the inside of his windowsill a few months previously, and a bit of a bad smell, but they had put it down to kitchen rubbish rotting during a summer heatwave.

Cause of death could not be established and there were no fingerprints, DNA or dental records on file for the occupant that might have been able to confirm that the body was his. His identity was accepted by the coroner on the basis of his ancestry and age. Sometimes, in the middle of a big city, surrounded by millions of people, you can simply be hidden in plain view.

With no clear starting point, and no immediate links to family, friends or colleagues who may be able to shed some light on what has happened to the deceased, no police investigation can truly get into its stride. In an ideal world, our police forces would have unrestricted budgets and unlimited personnel to devote to searching for missing persons and matching them with unidentified bodies. However, we are all aware of the restraints imposed by the real world and, given the ever-increasing numbers of people who go missing, there will be those who are never found, dead or alive. Virtually every UK police force holds human remains which, despite their best efforts, remain unidentified ('unidents'). Every year some of these will be buried without their given names, unbeknownst to their families and friends, because the investigative authorities have been unable to establish who they were in life.

When most of us die, our identity is not in question. The vast majority of us will do so under medical supervision at home or in hospitals, care homes or hospices. Those of us who die suddenly, for example as a result of an accident, will usually be carrying evidence of who we are, such as a wallet or handbag containing bank cards, driving licence or other documents bearing our name. Even when a body turns up out of the blue, most of the time it will be possible to name the deceased because they have died, say, in the house they occupied or the car they owned, and there will be clues in the paper trail that almost everyone leaves behind them. In such situations, the next of kin can be traced swiftly to verify their identity and assist with the investigation.

The biggest challenge is posed by a body found unexpectedly in an isolated place, maybe decomposed, and carrying no circumstantial evidence that could lead easily to their identification. There may be no hits, either, on DNA or fingerprint databases. This is when forensic anthropology comes into its own and offers the best and sometimes the only chance of reuniting the deceased with their identity in life.

The process we follow is well documented and involves a lot of common sense, logical scientific interpretation and attention to detail. As outlined in Chapter 2, there are two types of identity that we seek to secure when we are confronted with human remains: biological identity, which pertains to general classifiers, and personal identity, which should allow us to confirm the name of the deceased. One may lead to the other, but even when it does we have to be prepared for this to take a lot of time and patience. Obviously, we will process DNA and fingerprint information immediately in the hope that we might be able to skip straight to personal identification via a quick match. Often, though, that is a pipedream and we need to fall back on old-fashioned anthropological legwork.

Humans fall into several different general descriptor categories that can help to narrow the range of possibilities. The more recent the death, the more likely we are to be able to accurately determine the four basic components of biological identity: sex, age, stature and ancestry. These are the characteristics that allow us to put out a missing person notice to say that we have found the remains of a white female between the ages of twenty-five and thirty who was approximately 5ft 2ins in height. It is important that we get all of these broad-brush indicators right, as major mistakes may result in the deceased not being identified or a significant delay in the investigation. We may also end up in court as expert witnesses, so all of our opinions must be supported by sound scientific underpinnings and we must resist the temptation to stray into the realm of supposition.

◊

The determinant of the first component of our identity, sex, should be very straightforward as we expect it to be bimodal—male or female. The term 'sex' is used very specifically in our field and is not to be confused with 'gender': the former is used to denote the genetic construction of the individual while the latter relates to personal, social and cultural choices and may be at odds with our biological sex.

The genetic norm for the human genome is the presence of forty-six chromosomes in twenty-three pairs. One of each pair—half of your nuclear genetic composition—is donated by your mother and half by your father. Twenty-two of the pairs, while differing slightly from each other, have the same dual 'form' (a bit like pairs of vaguely matching black socks) while the twenty-third, the sex chromosomes, carry sex-related genetic information and are therefore quite distinct from each other (like mismatched socks of different colours).

As most of us will remember from our school biology lessons, the X chromosome contains the blueprint for 'femaleness' and the Y chromosome that for 'maleness' (specifically in the SRY gene of the chromosome). Females carry an XX combination of sex chromosomes and males an XY combination. So we all inherit an X chromosome from our mother.

If the chromosome donated by a father is also an X, the baby will carry an XX combination and develop into a female. When the father provides a Y chromosome, the baby will become a male. There are rare disorders of the sex chromosomes which produce alternatives to the usual pairing, including Klinefelter Syndrome (XXY) or Turner's Syndrome (XO), but these are so uncommon that I am not aware of ever having come across any of them in my entire career.

The developing embryo has a genetic sex from the moment the sperm fuses with the egg, but for the first few weeks of its development it appears to be asexual, with no obviously male or female external or internal characteristics. Even by eight weeks after fertilisation, there is little sign in the soft tissue of a human embryo to indicate whether it will become male or female, but by twelve weeks we start to see evidence of which sex the fetus may be. When Mum is having her first ultrasound scan, it might be possible to determine the sex of the fetus from the visible external genitalia.

These days, some hospitals choose not to confirm the sex of a baby, ostensibly because of staff shortages and the time required to make the assessment, but they will have other concerns, such as the risk of litigation if they get it wrong and the prevention of selective abortion by couples whose culture values one sex above the other. So the baby's sex may remain a mystery until the day it is born, as it always was in the days before ultrasound. I rather like the element of surprise and never wanted to know in advance what sex my babies were. As my father-in-law said, 'What does it matter as long as there is one head, ten fingers and ten toes?'

If the parents are keen to know, and the ultrasound operator is prepared to tell them, what he or she will be looking for on the image is the same visual external evidence used by the nurse, midwife or whoever is present to announce a baby's sex when it is born—if it has a penis, it is a boy, if it hasn't, it is a girl—which then becomes its legal description. There is much wrong with that as a basis for determining the legal sex of a child, biologically, socially and culturally. But since the dawn of time it is what we have relied on, and it is still the best we have.

From that moment on, the agenda for a baby's entire childhood is usually set in stone. They will be brought up as either a boy or a girl, with all the cultural trappings that definition brings, purely on the basis of whether or not they have a visible penis. If we got it right, when a boy goes through secondary sexual changes at puberty, he will be expected to develop appropriately sized external genitalia (testes and penis), male pattern hair distribution on his arms, legs, chest, armpits, pubic region and face, and a deepening of the voice. A girl will show breast development, widening of the hips, armpit and pubic hair growth and will begin to menstruate. Imagine the effect on your confidence in your own identity if, for the first twelve years of your life, you have believed yourself to be a boy and then you start to develop breasts, or, having always been told you were a girl, you notice hair appearing on your chest. Puberty is a period of great sensitivity and awkward awareness of our bodies at the best of times, and unanticipated alterations of such magnitude are understandably devastating to a young person.

In the vast majority of cases the designation of sex when we are born is correct but the forensic anthropologist must leave room for other possibilities. It would be handy for us if male skeletons were blue and female skeletons pink. Ridiculous as that sounds, let's use those colours for a moment to represent the maleness and femaleness of our bodies. Blue stands for the presence of the SRY gene and the production of the sex steroid testosterone, while pink represents its absence, which allows the other sex steroid, oestrogen, to predominate. Every baby has a combination of both sex steroids, just in different proportions. Since male embryos have one X chromosome, in addition to the dominant testosterone they are also producing oestrogen as a normal biochemical function. And females produce small levels of testosterone through anatomical routes that do not involve a Y chromosome, for example the ovaries and adrenal glands. If you are in any doubt, ladies, wait until after menopause, when oestrogen declines, allowing testosterone to assert itself, and watch your beard and moustache grow. The bearded lady beloved of Victorian circuses was not a freak of nature but a perfectly normal human variant.

Sex, or what we perceive to be maleness or femaleness, is largely about the interaction between genetics and biochemistry and the effects this has on all of the tissues of the body, including the brain. Imagine a genetically pink embryo that goes into overproduction of testosterone (as happens when a gene mutation causes adrenal hyperplasia), or a genetically blue embryo that either does not switch on its SRY gene or fails to produce enough testosterone (adrenal hypoplasia), or goes into overproduction of oestrogen, and you start to see how the interplay between genetic sex and physical appearance or psychological identity become conflated.

The forensic anthropologist needs to be aware that the features we see in the skeleton are a complex interaction between the genetic blueprint for sex and the effects of biochemistry, resulting in a grey area (or perhaps, given our colour scheme, that should be a mauve area) where genetically male individuals may display some feminine characteristics, and genetically female individuals appear more masculine, than their counterparts at opposite ends of the scale, where genetic sex and biochemistry are perhaps in closer harmony. What is so wonderful about the human is that we have so very many possible variations. It is what makes us truly fascinating to study as a species.

Even when human remains are relatively recent, determining biological sex can be challenging, especially if there has been some surgical intervention. It is therefore extremely important that we are not influenced by circumstantial evidence (the remnants of female underwear, for example) and that we are alert to the possibility of any congenital features or surgeries. The absence of a uterus may point to the remains being male, or they could be those of a woman who has had a hysterectomy or who was born without a womb due to agenesis (the failure of an organ to develop during embryonic growth). The absence of a penis or signs of breast augmentation may suggest that the deceased was female but equally, they could be indicative of elective transgender surgery.

After the Asian tsunami of 2004, in which a quarter of a million people lost their lives, the issue of biological sex and gender was prominent in the minds of many whose job it was to try to categorise and identify the dead. One of the affected countries, Thailand, is recognised as the transgender capital of the world. With a male-to-female operation here costing almost a quarter of the price charged in the US, over 300 such surgeries are performed in Thailand every year, and the third sex, or *kathoeys*, are acknowledged as a fully integrated sector of society. A strictly dimorphic approach to gender expression cannot be expected in this part of the world and in the wake of the disaster, external body assessments were always supported by internal examination.

Assigning biological sex becomes trickier as a body starts to decompose. External genitalia deteriorate quite rapidly after death and examination of internal anatomy through post-mortem dissection may be of limited assistance. Using DNA analysis to look for the SRY gene will help to confirm remains as male but it is of little value in verifying that they are female unless a full karyotype (a profile of an individual's chromosomes) can be established. So what do we do if all we have available to us is some dry, scattered or buried human bones?

In spite of my pie-in-the-sky wish for blue and pink bones,

an intact adult skeleton is actually already a reasonably reliable indicator of biological sex. The features we look for are those manifested when growth is accelerated during puberty in response to the effects of increased levels of circulating sex steroid hormones. If the predominant hormone is oestrogen, the changes in the skeleton will reflect what we read as 'feminisation' of the bones. It does not necessarily mean the individual is female, simply that it displays 'pink' characteristics. Here we expect the major change to relate to preparation of the pelvis to facilitate fetal growth and to permit a baby's head to pass through it unhindered.

Female pelves do not always comply with the norm, however. Cephalopelvic disproportion was a justified fear for pregnant women in the past. If the pelvis was not sufficiently capacious to allow the baby's head to enter, pass through and then exit the bony part of the birth canal, they might labour for days with no obvious or survivable solution to the impasse. Remember the fate of the Roman mother and triplets excavated in Baldock. Over the centuries, many have died from the traumas of childbirth.

In circumstances where saving a mother's life was considered more important than the survival of the baby, some gruesome obstetrical tools used to be employed to try to rescue her from cephalopelvic disproportion. The perforator, for example, was a metal implement shaped like a small lance that would be inserted into the mother's vagina and pushed beyond the cervix into the uterus, where it would 'perforate' the first part of the baby it encountered. As in most normal births this would be the head, the perforator was most frequently used to pierce the anterior fontanelle of the skull, the largest of the 'soft spots' that allow the bones to move in relation to each other so that the head can get through the birth canal.

A hook on the end of the perforator would then be moved around to find some bit of the skull it could latch on to, often an orbit (eye socket). In the process, it would disrupt some of the brain structure, making it easier for the baby's head to be forcibly pulled through the birth canal. Later perforators had a scissor-like action which enabled the baby to be removed literally one piece at a time.

Cephalopelvic disproportion is seen less often nowadays, mainly as a result of improved health but also perhaps as a rather brutal example of the survival of the fittest, whereby feto-maternal mortality has resulted in the phasing out of unsuccessful pelvic shapes. Even today, however, in some parts of the world, giving birth can be a hazardous business for both players who have a skin in the game. The World Health Organisation (WHO) reports an estimated 340,000 maternal deaths, 2.7 million stillbirths and 3.1 million neonatal deaths every year, almost all in impoverished countries. In sub-Saharan Africa, a woman's risk of dying while giving birth is 1 in 7 and cephalopelvic disproportion still accounts for over 8 per cent of maternal deaths.

Where there is access to adequate medical services, it no longer matters if the pelvis is the wrong shape or size because the baby can be removed by Caesarean section with a very high success rate for both mother and baby. In some more affluent countries, advances in anaesthesia and antibiotics have seen C-sections become almost elective as a result of the 'too posh to push' trend. And in situations where the wellbeing of both mum and baby are deemed too much of a financial risk for a hospital to follow the natural route, C-section is sometimes seen as a safer alternative.

So the twenty-first-century Western woman now comes in a range of shapes and sizes, all of which can legitimately be preserved in the genetic inheritance of pelvic shape. Ironically, it seems that determining sex from the pelvic bones may well be achieved with greater accuracy and reliability in archaeological specimens than in recent forensic samples, because the level of sexual dimorphism required to maintain successful childbearing heritage is being lost.

When the dominant circulating hormone is testosterone, its primary purpose during puberty is to increase muscle mass. We are all aware of how ingesting additional amounts of the male hormone in the form of anabolic steroids decreases fat levels and increases muscle bulk in bodybuilders. The bone-muscle equation is a simple one: stronger bones are required to withstand the forces exerted by the attachments of stronger muscles. In areas such as the skull, the long bones and the shoulder and pelvic girdles, we see more well-developed sites of muscle insertion. Testosterone, then, leads to masculinisation of the skeleton, but again, that does not necessarily mean that the remains are biologically or genetically male.

If there is no dominant circulating hormone, as will be the case in pre-pubescent children, the skeleton will tend to retain a paedomorphic or childlike appearance, which is generally interpreted as being more pink than blue. As the relevant changes we look for in the skeleton do not occur until puberty, sex cannot be determined with any degree of reliability from a child's skeleton.

If the entire adult skeleton is available for analysis, the forensic anthropologist will probably be able to correctly assign biological sex in about 95 per cent of cases, although different ancestral groups will show variations that we must take into account. For example, the Dutch are officially the tallest 'race' in the world but their babies are not any bigger than those of other Western populations. Not surprisingly, therefore, they have a very low level of obstetric complication because the proportionately larger Dutch female pelvis has not needed to adapt to ensure the safe passage of the baby. Women from other groups who are smaller in stature but deliver babies of the same size are believed to exhibit greater levels of sexual dimorphism in the pelvis as nature has sought to find a shape that will safely accommodate childbirth. This research tells us that it is potentially more challenging to distinguish between female and male Dutch pelves from skeletal remains.

Obviously, if the skeleton is damaged or disrupted, perhaps by fire or fragmentation, the determination of sex becomes increasingly difficult. To establish sex with some confidence requires us to be able to recognise the smallest fragment of bone and identify its location within the skeleton—is it from the distal humerus, proximal femur or a supraspinous piece of the scapula?—as we search for those independent areas that show the greatest differences between the sexes. So we are reliant on the survival of the most dimorphic areas of the skeleton. The shape of the greater sciatic notch in the pelvis, the prominence of the nuchal muscle markings at the back of the neck, the size of the mastoid process behind the ear and the presence of supra-orbital ridging under the eyebrows all hold important clues.

The greater the degree of sexual dimorphism, the more reliable the forensic anthropologist's findings will be when establishing sex from the bones. But we must always remember that the features on which we are basing our analysis are indicators of the extent and timing of biochemical influences, not proof in themselves of the biological or genetic sex of the individual.

Determining the sex of an unidentified body correctly is very important, because obviously, when trying to match a body to a missing person, being able to eliminate all members of the opposite sex will halve our pool of possible candidates. But the other side of the coin is the very real danger that, should we get it wrong, the chances of ever making a positive identification will be remote.

◊

Whereas we are quite successful at establishing the correct biological or genetic sex for an adult but pretty rubbish at it with children, when it comes to the second biological component of identity—age—it is the other way round. When you consider the trouble we have judging with any precision the age of living adults, who provide us with many more clues than the dead, it should not come as a surprise that determining age from remains is not easy, especially when they are skeletonised or, worse, fragmented.

In life, pinpointing age accurately becomes more difficult the older people become. We could walk into any primary school classroom and form a fairly good idea, to within a year or so, of the age of the children there. With a class of secondary school students, we are likely to get it about right when looking at them as a group, but there will be some individuals who appear either a lot older or a lot younger than the majority because they won't all be experiencing the various physical changes associated with puberty at the same time. As for guessing the ages of a roomful of adults, well, we all know that if we try that, we may flatter some but are likely to end up offending at least half of them.

In the early years of life there is a strong relationship between age, facial appearance and size. The face is a reliable indicator of age because of the way it needs to grow to accommodate dental development. I took a photograph every year of all of my children on their birthdays, which allowed me to create a chronological map of how and when their faces altered (all good scientists regard their children as their own personal little petri dishes). The first big change occurred in them all between four and five years of age, when the jaws, which form the bottom half of the face, have to grow sufficiently to allow the first permanent molar to erupt into the mouth when they are around six. The second significant change was just before puberty, as their jaws grew again to ensure that there was room for the arrival of their second permanent molars. Then all hell broke loose as the raging sea of hormones crashed into their lives (and ours) during puberty, when their beautiful, grown-up faces began to emerge.

The correlation between age and size in young children is reflected in the way we buy their clothes. These are sold by age rather than measurement because manufacturers can predict with some confidence that, say, between birth and six months, a baby will have a head-to-toe height of about 67cms (2ft 2ins). We don't look for a dress for a 3ft 5ins child, but for a four-year-old. The age range widens as the child grows: baby clothes are labelled in sizes of three-month intervals, then six months. Sizes for toddlers and upwards will rise in one- or two-year increments to the age of about twelve. When puberty strikes, changes to the body make the relationship between age and size far less predictable.

So when we are examining the remains of a fetus or baby, the length of the long bones within the upper limb (humerus, radius and ulna), and the lower limb (femur, tibia and fibula) will allow us to calculate its age to within a few weeks. With a young child we will be accurate to within a few months, and in an older child to within a range of two or three years.

There is a little more to it than simple measurements, however. In children, some bones are comprised of several parts, to allow for growth, which will eventually fuse on maturity. As the pattern of growth and fusion is closely related to age, the stage these bones have reached is a reliable guide. The adult human femur (thigh bone), for example, is a single bone but in children it consists of four different components: shaft, distal articular end (at the knee), a proximal articular head (at the hip) and the greater trochanter on the side of the bone where the muscles attach. The first part of the femur to convert from cartilage into bone is the shaft, which shows bone formation in the seventh week of intrauterine life. The centre of ossification (where bone first forms) at the knee will be visible around birth—indeed, its presence on an X-ray was used in the past as an indicator that a baby had reached full term and that the fetus was therefore considered to be clinically viable. This was important in the prosecution of mothers who concealed full-term births, which carried a harsher penalty than concealing the stillbirth of a fetus.

The bony part of the head of the femur, which forms the hip joint, begins to convert into bone by the end of the first year of life and the top of the greater trochanter, where the gluteus medius and gluteus minimus muscles, vastus lateralis and others attach, appears as a bony centre between the ages of two and five years. Then the different parts start to grow towards each other and eventually fuse.

Between twelve and sixteen years in females, and fourteen and nineteen in males, the head of the femur will fuse to the shaft, followed, within another year or so, by the greater trochanter. The last bit to fuse, to complete the adult bone, is the distal end, at the knee, which occurs between sixteen and eighteen years in girls and between eighteen and twenty in boys. When all the components have fused together there will be no more growth in that bone. When all of the bones finish growing in length, we have reached our maximum height.

The growth and maturity of most bones in the developing skeleton follow a pattern that allows us to estimate a likely age, providing, of course, growth is proceeding as we expect. Some parts of the body offer more information than others. An adult hand, for example, has around twenty-seven bones, whereas in a child of ten these will be made up of at least forty-five separate parts. This makes it a good witness in establishing age in life as well as in death. As it is also easily accessible, and the most ethically acceptable part of the body to expose to the ionising radiation of X-ray, it is often used to determine whether someone presenting themselves as a juvenile for immigration or refugee purposes really is a child.

Over half of the world's population is born without a birth certificate and therefore no documentary proof of their age. This causes little problem when people remain within a geographical area where the powers that be accept it as commonplace, but when someone who does not have such paperwork migrates to a country where the fabric of society is dependent on official evidence of identity, they can come into conflict with the authorities.

The countries who have signed up to the United Nations Convention on the Rights of the Child agree to protect children from harm, to house them, clothe them, feed them and educate them. When potentially bogus immigrant claimants, or children who have slipped through the net, are identified by the authorities, forensic anthropologists are sometimes asked to assess their age, especially if they come to the attention of the criminal court as either a perpetrator of an offence or a victim, such as a child suspected of having been trafficked.

My colleague Dr Lucina Hackman is one of only two practitioners qualified in the UK to perform age assessment in the living. She uses medical imaging of the skeleton—CT scans, X-rays or MRI—to determine an approximate age which may be brought before the courts as confirmation of the age of criminal responsibility or consent, or as evidence in cases dealing with the international rights of a child.

Once an individual is beyond childhood and adolescence, there is a weaker correlation between age-related features and actual chronological age. We can be reasonably accurate to within five years with people up to about forty years old, but after that changes in the human skeleton are largely degenerative and, to be honest, we all fall apart at different rates, depending on our genes, our lifestyle and our health. We probably all know a sixty-year-old who looks forty, and vice versa. In looking at remains of individuals in their fifth and sixth decades, we tend to resort to descriptions like 'middle-aged adult'—I hate that label, especially when it defines me—and when dealing with anyone over about sixty we talk about 'elderly adults'.

Outrageous! It just goes to show how bad we are at assigning age with confidence at the upper end of the scale, whether to a living individual, a dead body or skeletal remains.

So we are adept at determining sex in the adult, less so in the juvenile; quite impressive at establishing the age of a child but mediocre when it comes to grown-ups. What about the other two biological categories, stature and ancestry? One we are really good at and the other pretty poor, by and large. Ideally, the assessment we are best at would be of the most value in establishing the identity of the deceased. Oh, if only nature were so kind! Unfortunately, the one we are really good at, stature estimation, is probably the least important of all four of the biological characteristics.

This isn't exactly showing the glories of forensic anthropology in the way the television shows do, is it? But in the real world it is important to recognise that if the identity of an individual is not easy to determine, the solution will come down to experience, expertise and probability across all four identifiers. The anthropologist who asserts absolute confidence in the sex, age, stature and ancestry of a skeleton is a dangerous and inexperienced scientist who doesn't understand human variability.

◊

In the UK, the height of most adult individuals falls within a range of sixteen inches, between 5ft and 6ft 4ins (1.5m to 1.93m). Anyone outside that bracket might be considered unusually short or unusually tall. The average height for a female is 5ft 5ins (1.65m) and for a male 5ft 10ins (1.78m). Of course, stature is strongly influenced by genetics and environment. If you have tall parents, you will probably be tall, and if they are short you are likely to be short, too. We can predict the adult height of a child either by doubling their stature at the age of two (isn't it incredible that we grow to half our full adult height within our first two years?) or by calculating what is called MPH (mid-parental height). For a boy, in centimetres, the equation is: father's height + mother's height + 13 ÷ 2; for a girl: father's height—13 + mother's height ÷ 2.

To illustrate the influence of genetics we need only look at the variations in average height in different parts of the world. The tallest men are the Dutch, who are on average 6ft (1.83m), and the shortest are from East Timor, at 5ft 3ins (1.6m). Latvia has the tallest women, pipping the Dutch ladies to the post at 5ft 7ins (1.70m), and Guatemala the shortest at 4ft 11ins (1.5m).

The tallest person ever recorded for whose height there is reliable proof was Robert Pershing Wadlow from Illinois in the US, who was 8ft 11ins (2.72m) at the time of his early death at twenty-two. He unfortunately suffered from an excess of human growth hormone, and was still growing when he died in 1940. The record-holder at the opposite end of the scale is Chandra Bahadur Dangi from Nepal, at 1ft 9½ins (54.6cm), a primordial dwarf who enjoyed a long life for those with his condition—he died in 2015 at the age of seventy-five.

As is demonstrated by these examples, genetics is not the only influence on our adult stature. As well as the rarer impact of growth disorders, more commonplace factors such as nutrition, altitude, disease burden, growth variations, alcohol, nicotine, birth weight and hormones will all affect how tall we will be as an adult. With fully favourable conditions, a child will reach their height potential. If they experience adverse conditions in their first fifteen years or so, they are likely to be shorter than expected.

As Western culture views tallness as desirable and shortness as a disadvantage, most of us have a tendency to overestimate our own height. And when we estimate the height of others, we base our assessment on our perception of our own height and so overestimate theirs, too. Unwilling to acknowledge that we shrink with age, we continue through our lives to claim the height we were in our prime, even though we become shorter whether we like it or not. Once we pass forty we lose about a centimetre every decade and, after seventy, a further three to eight centimetres.

Our height is made up of the length and thickness of all of the components of our body, from the skin on the soles of our feet to the skin on the top of our heads, encompassing bone heights and lengths (calcaneus, talus, tibia, femur, pelvis, sacrum, twenty-four vertebrae and skull), plus joint spaces between these bones and also the thickness of the cartilage between the bones of each joint. With age, cartilage thins and some of the joint spaces collapse. Clinical conditions such as arthritis and osteoporosis will also alter the bones and joints and reduce overall height. And believe it or not, our height varies according to the time of day: we are on average half an inch shorter by the time we go to bed than we were when we got up. We lose most of that height within three hours of rising, as our cartilages settle and compress and decrease our joint spaces.

It would be quite a challenge if, when trying to determine the height of an individual from a skeleton, we were to attempt to add together the measurements of all the different bones, cartilage and spaces that contribute to it. When a body is found with all the bones pretty much where they should be, there will be a lot of soft tissue still present, so we will get out our tape measure and record the recumbent stature there and then. In the mortuary, we will follow the same procedures as we do to calculate the age of a child from their long bones. It stands to reason that if you have long arms, and especially if you have long legs, you will be tall and the opposite is also true. We measure the length of each of the twelve long bones (the femora, tibiae, fibulae, humeri, radii and ulnae, of which we have two apiece) on a device called an osteometric board and place the values into appropriate statistical regression formulae for the sex and ancestry of the individual. The resulting living stature will be estimated to within about three or four centimetres of the person's living height.

The reality is, though, that in a forensic investigation, stature is unlikely to be a major identifying feature in its own right unless you are exceptionally tall or small. I have known of families who have latched on to the vain hope that remains we have examined are not those of their son, even questioning a DNA result confirming his identity, purely because they have been given a most likely height of around 5ft 6ins and he was actually 5ft 7½ins. That is why we provide the full margin of error and suggest a range.

Our fourth biological identifier is ancestry, or what may in the past have been called 'race'. We now avoid this more emotive term because of its negative associations with social inequalities, and the risks such connotations carry of preconceptions and misconceptions, and also because the biological evidence for which we are looking is of a long-past origin. Assignation of ancestry is potentially of enormous interest to the investigative process but often forensic anthropologists are not talking the same language as the police. What the police will want to know is whether they should speak to members of, say, a Polish or a Chinese community. Unfortunately, we are unable to distinguish between groups such as these and others of close biological proximity just from looking at skeletal remains.

We categorise people on the basis of a variety of physical traits: the colour of their skin, their hair or their eyes, the shape of the nose or eyes, the type of hair they have or their language. Clusters of multi-locus genetic data have largely confirmed the accepted premise that, notwithstanding a bleed of features across geographical regions, we used to be able to genetically separate the world into four basic ancestral origins. The 'out of Africa' concept, which classified the first ancestral group as people originating from sub-Saharan Africa, still holds firm. The second group stretched from North Africa across Europe and east to the border with China. The third encompassed the eastern regions of the Asiatic land mass and, across the North Pacific Ocean, the North and South Americas and Greenland. The fourth, more geographically isolated, region consisted of the South Pacific Islands, Australia and New Zealand. This resulted in the four archaic classifications of Negroid, Caucasoid, Mongoloid and Australoid.

While we may quite easily categorise the origins of our ancestors, things have become trickier in the more recent past and I suspect many of us would get a few surprises regarding our own histories if we investigated our genes. In our palaeo-distant pasts, it is likely that cross-linkage between these groups was limited, but in our smaller, modern world, where interactions have been more common and more frequent over the generations, the genetic signal for each of the four discrete classifications is becoming ever fainter.

What genetics cannot tell us with any reliability is the difference between a man from China and a man from Korea, or a British woman and one from Germany. Therefore it is of no value whatsoever in assisting us with assessing the ancestry of a person descended, for example, from an Indian maternal grandfather, English maternal grandmother, Nigerian paternal grandfather and Japanese paternal grandmother.

There are some basic characteristic differences, particularly in the facial region of the skull, that can be detected in individuals in whom the manifestations of ancestral origin are more protected. We do have computer-based systems that can process a variety of skull measurements and give us suggestions as to the most likely ancestry for an individual, but these must be viewed with some caution. What we will be hoping in such circumstances is that hair or other soft tissue survives that can assist us, or that clues may be provided by personal effects such as clothing, documents or religious jewellery. DNA analysis is our best chance but it will only tell us about ancestral origin, not nationality. It cannot tell us whether a person of Indian ancestry was born in Mumbai or London. Only stable isotope analysis offers us any help here.

◊

Once the four biological parameters have been determined, our next job is to find individual identifiers that will enable us to focus on a single person to the virtual exclusion of all others, using one or all of the INTERPOL-approved primary methods: DNA comparison, dental records or fingerprints. Fingerprints are unlikely to be obtainable from skeletal remains but it is sometimes possible to retrieve them from even quite badly decomposed bodies.

If DNA databases provide no matches, the police may go public with the information to try to generate leads. To establish the personal identity of the deceased—their name—we need intelligence to follow, and it is hoped that the community will respond with suggestions that can be eliminated or pursued further by investigators. When the police put out the message that the deceased is, say, a male aged between thirty and forty, around 5ft 8ins in height and black, they are obviously narrowing the range of possibilities by ruling out females, children, old people, the short, the tall and the other umbrella ancestral groups. But there will, as discussed previously, still be thousands of individuals registered as missing who will fit that broad biological profile.

Missing person posters circulated by the police will probably include an image of what the person may have looked like based on a reconstruction of their face, like the one that helped to identify our woodland suicide in Chapter 2. The forensic artist or reconstruction expert will rely on the biological characteristics that we have provided being correct. If we say the body is that of a woman when it is a man, or we say they are black when they are white, or around twenty years old when they are more like fifty, then the reconstructed face is never going to resemble the person it should depict.

An illustration of how dramatically these reconstructions can help to accelerate identification is provided by a 2013 Edinburgh case in which dismembered female remains were found in a shallow grave on Corstorphine Hill. The only clues were some distinctive rings on the fingers and some extensive dental work. A likeness produced by my craniofacial colleague Professor Caroline Wilkinson and circulated internationally was recognised by a relative in Ireland as Phyllis Dunleavy from Dublin. Mrs Dunleavy had been in Edinburgh staying with her son, who had claimed she had returned to Ireland. The identification led to him being charged with her murder within a month of the discovery of the body, and subsequently convicted.

The shorter the time between a death, the deposition of remains and the identification of the deceased, the greater the potential for retrieval of evidence will be. In this case, the speed with which the body was identified unquestionably enhanced the investigative process and was key to the success of the prosecution.

When we have a potential identity for our body, DNA extracted from the bone can be compared with a sample from a mother, father, sister, brother or children. It may even be possible to acquire the missing person's own source DNA, perhaps from a toothbrush, hairbrush or ponytail bobble that still holds some shed hair with cells at the base. In the UK, we may be able to access source DNA from the Guthrie cards retained by the National Health Service. These contain small blood samples taken from almost every baby born in the UK since the 1950s, obtained by a pinprick test on the heel, blotted on to paper and used to test for a variety of genetic conditions including sickle-cell disease, phenylketonuria, hypothyroidism and cystic fibrosis. Almost all NHS authorities keep these Guthrie tests, although their use for the purpose of forensic identification is somewhat contentious as it is not one for which the original consent was given. It was a Guthrie card that provided the match for at least one individual who died in the Asian tsunami of 2004, enabling identity to be confirmed and the body to be returned to the family. The privacy question, and whether the outcome of a positive, or indeed negative, identification can ever justify the absence of consent, is one for the lawyers to debate.

The UK National Criminal Intelligence DNA Database (NDNAD), set up in 1995, is the largest national DNA database in the world. Over 6 million profiles are stored there, representing nearly 10 per cent of the country's population. Approximately 80 per cent are from men. Recent statistics suggest that the database helps to identify a suspect in around 60 per cent of criminal cases. A full national database of DNA for all UK citizens would be relatively straightforward to introduce and would very probably reduce the number of unidentified bodies and unsolved crimes. However, opinion is strongly divided on whether the benefits of a national system should be permitted to outweigh the right to privacy and anonymity. That is one very angry hornet's nest and I suspect the arguments will roll on for a very long time.

Sometimes, often in cold cases, an individual's DNA sample can result in them unwittingly helping to bring to justice an offender to whom they are related. One example is the case of the shoe rapist, who raped at least four women in south Yorkshire in the 1980s and attempted to do the same to two others. After attacking them he would steal their shoes. Some twenty years later, the DNA sample of a woman charged with a drink-driving offence was placed on the DNA database and a familial genetic link to the rapist was highlighted. She was his sister. When the man's workplace was raided, police found more than a hundred pairs of ladies' shoes, including those belonging to the rape victims. He was given an indeterminate sentence and the judge ordered that he should serve at least fifteen years in prison.

Although there is no central database for dental records, most UK citizens visit a dentist at some point in their life and so there will be evidence of the work undertaken on their teeth—if we can track down their dentist. Many people have more than one set of dental records. Not everybody stays with same dentist and, with many procedures not available on the NHS, records of private treatment may be held by a different dentist from the one with whom we are officially registered. As increasing numbers of patients choose to go abroad for better, and often cheaper, cosmetic alterations, some of these won't even be in the same country. Such records are rarely traceable. And given that many dentists only keep documentation for audit purposes, the information available may be hard to interpret in relation to the mouth that is being investigated.

A recent additional complication arises, ironically, from advances in dentistry. In common with others of my generation, I do not have a straight tooth in my head. I have a palate typical of my northern European ancestry which is not wide enough for all my teeth, so they are all crowded in, higgledy-piggledy, like headstones in an old graveyard. By the time I was fourteen, I didn't have a single tooth that wasn't filled, either. My trace metal levels of silver, mercury, tin and copper are probably off the charts. God bless the good old Scottish diet and the lack of fluoride in our water. The result is that, although my mouth may not be pretty, it is highly unlikely to look like anybody else's, especially after root-canal work, veneers and wisdom-tooth extractions over the years. If my body needed to be identified, my dentist would be able to confirm without hesitation that it was mine.

By contrast, many of today's teenagers have perfect teeth. Their dental braces have ensured that every tooth is straight so that they can flash their bleached Hollywood smiles (teeth are supposed to be slightly yellow, not white), and if they have any fillings at all, these are likely to be white as well and really difficult to spot. I'm sure our family dentist would have trouble confirming the identity of my daughters from their dentition.

In the UK, fingerprints can be taken from anyone who has been arrested or detained on suspicion of committing, or having committed, an offence. Ident1, the searchable database for finger and palm prints, holds over 7 million ten-prints (prints from each of the ten digits) and makes more than 85,000 matches every year with evidence recovered from crime scenes. This system is also used at border controls, where it is estimated that over 40,000 identities every single week are checked by UK visa and immigration officers.

As long as we have biological parameters and a likely identity for remains, one or a combination of these three INTERPOL-approved identifiers will usually enable us to confirm personal identity. Even when primary identifiers cannot assist, often there are secondary sources such as scars, tattoos, clothing, photos or other personal effects which allow us to be reasonably certain that the deceased is a match for a specific missing person.

The bodies we cannot name—like the names of the missing whose bodies have never been found—are those that haunt forensic anthropologists. None more so, for me, than that of a young man whose body was discovered at Balmore in East Dunbartonshire. Further information about him is listed at the end of this book. In an absolutely unashamed appeal for assistance, I would like to describe him here because there may be someone out there who can help us solve the mystery of his identity and allow us to give him back to his family.

The story began for us when our team at Dundee University was contacted in January 2013 regarding some badly decomposed human remains found hanging in a secluded area of woodland in the Balmore area. They had probably been there for between six and nine months when they were discovered on 16 October 2011. Missing person checks and DNA profiling had yielded no matches. The personal effects associated with the remains had revealed nothing that could assist with identification. The procurator fiscal was not of the opinion that the death was suspicious, believing it to have been suicide, but had requested that one last attempt be made to secure identity before the body was buried as an 'unident'. We were asked to examine the skeletal remains with a view to formulating a biological profile (Dr Craig Cunningham), reconstructing the face (Dr Chris Rynn) and analysing the personal effects found with the remains (Dr Jan Bikker).

The pelvis, skull and long bones all told us that they were likely to be those of a male. His age was put at between twenty-five and thirty-four, indicated in particular by the ageing of his costal cartilages (the soft tissue that connects the rib end to the breast bone), his pubic symphysis (the joint between the right and left halves of the pelvis at the front, behind the pubic region) and the junction between his first and second sacral vertebrae (at the base of the spine). He was probably of northern European ancestry and had fair hair—some of which was still present. His height was between 5ft 8ins and 6ft 1ins (1.75m and 1.85m) and he was of slight frame. No fingerprints were obtainable. He had previously had some dental work but this could not be traced without a possible name.

What was perhaps the best chance of identifying him lay in his many injuries. A healed fracture to the left nasal bone may have been visible in life as a crooked nose. There was a healed fracture to a bone at the base of the skull called the lateral pterygoid plate. Both were likely to have resulted from the same traumatic event, probably several months before his death. Had he had an accident, or had he been the victim of a serious beating?

An additional fracture to the left side of his jaw, which had been missed in the original postmortem examination, had not healed successfully but may have been sustained at the same time. This should have been treated in hospital with plates and screws but since it had not, he would have experienced tremendous pain every time he tried to eat. Did this relentless agony lead to a decision to commit suicide?

His kneecaps showed some evidence of articular degeneration, unusual in someone so young, and it is likely that he also found walking painful, so perhaps he limped. His upper left central tooth was fractured, possibly in the same incident that caused the injuries to the rest of his face, and the chip would have been visible every time he opened his mouth.

He was wearing a light blue, short-sleeved, V-neck polo shirt with a white printed design on the front featuring text and stamps; a long-sleeved, dark blue, crew-neck cardigan with a front zip; button-fly denim jeans and laced black and grey trainers with a red sole. The length of the jeans was in keeping with the calculated stature range and the waist size corresponded with the small size of the polo shirt and cardigan. Do these clothes ring any bells? Please do look at the list at the end of the book for further details, including brands, logos and precise sizing.

Who was this man? One suggestion was that he may have been an apparently homeless chap said to have been living rough in the woods around Balmore. He fitted our description and I understand that he has not been seen since, so he remains a possibility. But as the police had no name for him, the trail went cold.

Maybe the man from Balmore didn't want to be found. Perhaps he was scared and in hiding. Who was responsible for his broken jaw? Why did he choose to live with his pain and distress rather than seek medical assistance? Why did he take his own life? What a strange phrase that is. In what way did he 'take' it? From whom did he 'take' it? Our language around death can be inconstant and nebulous. She throws up so many questions for us and sometimes we simply cannot find the answers on our own.

I believe that if you have a right to an identity in life, you have the same right to it in death. A few may choose not to exercise that right, but those of us who are left behind have a duty to try to restore it to a person who has been deprived of it if we possibly can. The passage of time does not alter that. It makes the task increasingly difficult, but cases such as that of Alexander Fallon, who was finally identified sixteen years after he died in the 1987 King's Cross fire, demonstrate that it is still achievable.

Somewhere there must be a family who is missing the man from Balmore. It is our fervent wish to be able to give him back to them.

CHAPTER 9

The body mutilated

'Let fire and cross, flocks o' beasts,
broken bones and dismemberment come upon me'

Ignatius of Antioch, bishop and martyr (*circa* 35–107)

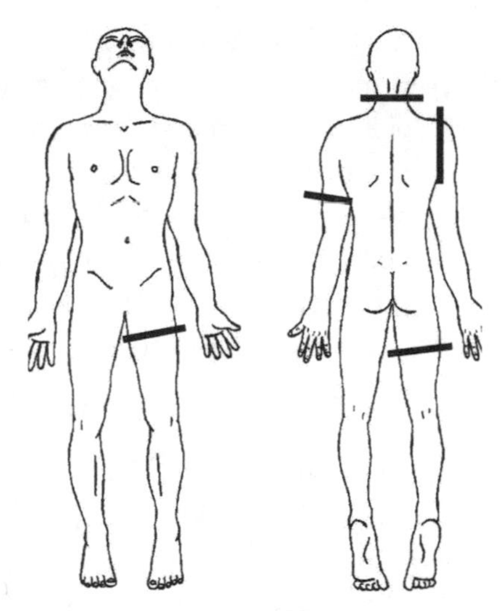

The position of the dismemberment cuts on Gemma McCluskie.

The act of separating a body into parts as a sacrifice or punishment is present to some extent in almost all cultures. Wood etchings depicting the Spanish atrocities in the New World or the eighteenth-century satirical engravings of the Day of Judgement made by anatomist William Hunter all convey a human acceptance of the practice of deconstructing the corporeal whole. And indeed it has been performed in assorted ways in almost all societies at some stage of their history for a variety of cultural, religious and ritualistic reasons. It is only in relatively recent times that desecration of the human form by dismemberment has come to be seen as repugnant and synonymous with criminality, usually murder.

Of course, not all dismemberments are criminal. An accident at work or a sporting misadventure may lead to the loss of a limb, and suicide by jumping in front of a train can cause extensive corporeal dismemberment and wide dispersal, as can violent mass fatalities such as plane crashes, resulting in detached body parts being found or having to be searched for.

Of the 500 to 600 murders a year that take place in the UK—fewer than 1 in every 100,000 members of the population—approximately three are recorded as involving criminal dismemberment, so it is certainly not common. But when it does happen, it fires the imagination of the public and the media and tends to garner more column inches than almost any other kind of crime, providing a rich seam of inspiration for novels, television dramas and horror films.

In the real world, how do you dispose of a body so that no one will ever find it? Everyone thinks they have an answer to that question (many of them informed by watching *Dexter* on

TV) and a theory on what constitutes the perfect murder. But of course, if a murder is perfect, no body will ever be found and no perpetrator punished—the only crimes we hear about are those that are imperfect. If a killer has got away with it, which has certainly happened, we remain in blissful ignorance of how they did it. Even when no body comes to light, prosecutions do, of course, take place, though such cases are more difficult to prove.

A body is a very unwieldy object to handle at the best of times and its size, weight and inability to co-operate can make its disposal somewhat troublesome for anyone trying to conceal a death. Unless the remains are going to stay within the premises where the death has occurred (and bodies do get found under beds, in cupboards and wardrobes, behind bath panels, in attics, basements, gardens, sheds, garages, up chimneys and under new patios and driveways), it will need to be transported elsewhere. Indeed, there is often an urgent requirement to remove it from the scene and for the murderer to literally distance him or herself from the evidence.

There are, however, many practical problems to be addressed. Can you safely move it intact? If not, where are you going to cut it up? What are you going to use? What are you going to wrap the pieces in? Because they are going to leak, believe me. What type of receptacle will be big enough? When do you move it? Are you likely to be seen? There may be CCTV cameras everywhere or passersby who might notice you. What kind of transport are you going to use? Where are you going to take it? How are you going to dispose of it when you get there? Can you do it on your own?

If a murder has been premeditated, the killer might have considered in advance what to do with the body, but as most killings occur in the heat of the moment there is usually no forward planning. Once the perpetrator realises the victim is dead, whether or not it was intentional, all these questions and more are likely to flood into an already panicked mind. As a result, the solution arrived at is often poorly thought through and executed on the spur of the moment. Very few people will have experience of this situation. For most, it will be the first and only time they take a life and dismember a body, and so they tend to unwittingly leave a trail of evidence for both police and scientific investigators.

Whether or not a murder has been planned is important as premeditation carries a higher sentencing tariff if the court finds the offender guilty, as does intentional dismemberment of a body. If murder is regarded as perhaps the most heinous of all crimes, the deliberate desecration of remains is seen as an additional insult, a step beyond the boundaries of humanity. Proof of dismemberment is therefore treated as aggravation of a homicide and punished accordingly. The fact that those incarcerated in Her Majesty's prisons on whole life tariffs are all there for murder and aggravated murder demonstrates how seriously society takes these crimes.

As criminal dismemberment cases are so out of the ordinary, police officers might only ever be faced with one, if any, in their entire career. So they often seek advice from other professionals, including forensic pathologists and anthropologists, with more experience in this area. My team at Dundee sees enough of these cases to warrant being designated UK experts by the National Crime Agency.

There are five main classifications of criminal dismemberment, based essentially on the intent of the perpetrator. Defensive dismemberment is by far the most common and occurs in about 85 per cent of cases. This odd term reflects the functional requirement to get rid of a body as quickly and conveniently as possible. The motive is the elimination of evidence and concealment of the offence—which is usually, but not exclusively, murder. In other words, it is a means to an end, as opposed to an element of the original crime. It appears logical at the time to reduce a body to pieces of manageable size that can be removed from the scene efficiently and disposed of without drawing attention to the death.

Statistics tell us that most killers and dismemberers are known to their victims and that the murder is most likely to occur in the home of either the victim or the offender. The dismemberment usually takes place at the murder site, using tools generally available in our kitchens, sheds and garages. Not surprisingly, the bathroom—designed to deal with a lot of fluid and to be easily wiped down and cleaned—is the most frequently chosen domestic site. It also has, in the bath or shower, a receptacle specifically tailored to the size and shape of the human body. So in a case of suspected criminal dismemberment, most scene-of-crime officers (SOCOs) will start their investigation in the bathroom.

Bending down to saw or hack a body in the confined space of a bath or shower is tricky, so spatters of blood and body tissue are common and, however thoroughly the perpetrator believes they have cleaned up afterwards, swabs taken around the walls, the base of the taps and the floor will often reveal traces of blood. Investigating the contents of the U-bend can also be productive, as is looking closely at the surface of the bath or shower for marks left behind by a saw or cleaver. It is hard to cut up a body without the blade catching somewhere.

Defensive dismemberment is generally characterised by an anatomical approach to the process since a body is easiest to handle when divided into six parts: head, torso, two upper limbs and two lower limbs. An intact torso is still a very heavy and bulky section to shift, but bisection tends to be avoided due to the complications of exposing and cutting the viscera. Cutting through bone isn't straightforward, either, because it is so hard—in life it must be strong enough to support our body weight day in, day out and to withstand knocks and falls. Knives will not do it. Mostly implements such as saws, cleavers and even garden loppers are used. Limbs are often the first parts to be removed. As they are connected only at one end, they get in the way and the axis of the body will be more manageable without them hanging around. Usually the cut is attempted across the single bone of the thigh (femur) or arm (humerus) to sever the limbs from the main trunk.

The head is more problematic to remove as the neck is comprised of a series of interlocking and overlapping bones, a bit like child's building blocks, making a clean cut difficult to achieve. The real challenge here, though, is a psychological one. Most perpetrators will inflict this ultimate insult with the body lying prone (face down) rather than supine (face up): it is assumed that having to look at the victim's eyes may deter them from decapitating a supine body.

In terms of the practicalities, dismemberers probably decide that it will be easier to remove the head from the back whereas, if you know what you are doing, it is in fact much easier to do it from the front. The prospect of removing the head at all proves too daunting for many, and this may be the point at which bisecting the torso, difficult and unpleasant as that may be, begins to seem preferable. Going for that option is usually a big mistake as the ensuing mess is much more challenging to conceal. While the torso is intact, the internal organs will remain contained within the body cavities, but once they are exposed they will leak copiously and create a really noxious stench.

Unless the body parts are going to be hidden in the dwelling, they must then be removed from the bathroom and taken out of the premises. Most perpetrators will wrap the pieces in plastic bags or bin liners and clingfilm is also sometimes used, as are other forms of household plastics or linens like shower curtains, towels or duvets. In the process, the killer has to try not to get blood or tissue on the outside of the wrapping or the container in which the body parts will be carried. Bin bags, for example, can be torn by the sharp ends of cut bone and towels soaked with sufficient volumes of blood will leak.

The notion of the body in the rolled-up carpet generally belongs to the days of Ealing comedies—the most common conveyances now are wheeled suitcases or rucksacks. Nobody looks twice at a person lifting a suitcase into a car or a taxi, or towing it down the road on foot. Offenders then tend to head for somewhere familiar to them to get rid of the remains. Water is the most common choice: rivers, lochs, lakes, ponds, canals or the sea.

Defensive dismemberment also covers situations where the intention is to obscure the identity of the deceased. In these cases, dismemberment will be anatomically focused. The usual targets are the face (to obfuscate visual identification), teeth (to prevent comparison with dental records) and hands (to destroy fingerprint evidence). Sometimes skin may be cut off to eradicate evidence of tattoos and body parts stripped of all recognisable jewellery.

Fortunately, these efforts tend to be unsuccessful. Killers may think they know the areas of the body that are key to identifying remains, but the reach of forensic science is greater than many of them imagine. As we have seen, there is almost no part of the body that cannot be used in some way to assist with identification. And over the last generation, the rise of a culture of artistic experimentation on the canvas of the human form has provided forensic experts with a wider range of potential leads. Increasing numbers of us are having ourselves tattooed, or our skin pierced in all kinds of places, or our breasts, buttocks, pecs and even our calves sculpted with silicone implants—all of these personal modifications provide new opportunities for identification, as long as sufficient evidence of the alteration survives.

Obviously, the recovery of an intact body offers the best chance of success but sometimes badly decomposed or even dismembered remains can still provide important clues to identity. A perpetrator may remove jewellery from piercings, but if the skin is still there, the presence of the puncture marks have a value. The silicone remnants of implants—and if we are really lucky, a visible and still decipherable batch number—can be most helpful in tracking down where surgery has been performed and upon whom. A tattoo might be removed through skinning or dismemberment, but if you understand how tattooing works, it takes only a little bit of anatomical know-how to uncover some evidence of inking.

The skin has three layers: the epidermis, dermis and hypodermis. The outer layer, the epidermis (the bit we can see), is comprised of dead cells that slough off continually, at a rate of almost 40,000 a day. So any ink laid down here will fade and eventually be lost, as happens when you have a temporary tattoo, such as a henna inking. Beneath the epidermis lies the dermis, the layer tattoo artists aim for with their needles. This has lots of nerve endings but no blood vessels, which is why having a tattoo is going to be painful, but it shouldn't bleed. Think about how a paper cut, which doesn't always even draw blood, can hurt like the devil. This is because you have cut through the epidermis and into the dermis, with its sensitive nerve endings, but not deep enough to reach the blood vessels in the hypodermis.

Tattooing too far into the hypodermis is futile as the cardiovascular system will remove the ink as waste material and the body will excrete it. The molecules of the dyes used in tattooing are large and designed to be inert so that they are generally not broken down by the body, do not interact with the immune system and can remain successfully trapped in the dermal layer between the epidermis and the hypodermis—a bit like the cheese in a sandwich. Inevitably, some breakdown of the ink molecules will occur (tattoos do fade with time) and these remnants will be vacuumed up into the lymphatic system for disposal.

Each of the lymph vessels within the dermis will eventually connect into a terminal swelling. We have many of these lymph nodes scattered throughout our bodies, but there is a high concentration of them at the top of our limbs, in the groin and armpit in particular. At these sites they act a bit like a sink trap in a shower that collects hair: as the ink molecules are too big to pass through the nodes, the dye accumulates there. Which is why, in people with tattoos, the nodes eventually take on all the colours of the inks.

We have always been aware in anatomy of this probably harmless quirk: as a student, when dissecting the armpit of dear Henry, my cadaver-teacher, who had those old-fashioned sailor's blue anchor tattoos on his forearms, I noted that his lymph nodes were blue with little hints of red from the lettering associated with the image. Today, with tattoos becoming a must-have fashion accessory (in the US, nearly 40 per cent of young people between the ages of twenty and thirty have at least one), we see this far more often and, reflecting the rainbow of colours used by tattooists, the current population's lymph nodes are truly spectacular in their kaleidoscopic variation.

Imagine a dismembered torso is found, with no trace of the upper limbs. If it is still sufficiently fleshed, we can look for the lymph nodes in the armpits, analyse any dyes found there and they will tell us whether tattoos have been present on one upper limb or two, and what colour the tattoos were on those missing limbs. Unfortunately, we can't predict whether the image might have been a dolphin, some barbed wire or simply the word 'Mum'. But when there is next to nothing to go on, it's a start.

Much to my chagrin, one of my daughters has three tattoos (that I'm aware of), visible piercings and very probably other modifications that a mother should never know about. Even I might consider a modest tattoo one day, if only for practical reasons. I have toyed with the idea of having the words

'UK, Black' and my national insurance number inked under my watch strap, hidden discreetly from view, after the fashion of Lady Randolph Churchill, who was reputed to have had a tattoo of a serpent around her wrist. Then, if I should be involved in a mass fatality, or my remains are not found for some time after I die, that strip of skin will give the identification team a head start and make their job just a little bit easier. I haven't yet plucked up the necessary courage. Remembering how anxious I was when I went into our local jeweller's in Inverness on my fifteenth birthday to give myself a present of pierced ears lobes—I told them that if they asked me to make an appointment and come back, I would lose my nerve, so it had to be now or never—I wonder whether perhaps a tattoo might be just a step too far for me.

Some defensive dismemberers may make attempts to destroy remains completely, for example through chemical treatment or burning. Dissolving a body is not as straightforward as some people think. Strong acids or alkalis are dangerous liquids to work with and obtaining them in sufficient quantity will arouse suspicion. Finding a container they won't corrode in the process would not be easy, either.

I once worked on a case in the north of England where a man admitted to murdering his mother-in-law and disposing of her remains. His claim that he dissolved her in a bath of vinegar and caustic soda and her liquefied body disappeared down the plughole was rather let down by his dodgy grasp of chemistry. Vinegar being an acid, and caustic soda an alkali, they would balance and cancel each other out. Furthermore, there is no way that chemicals sold over the counter would be strong enough to dissolve adult human bones, teeth and cartilage into a liquid that you could simply pour down the drain. The acid would have to be super-strong, and the chances of domestic plumbing coping with it are close to zero.

Even when a confession is forthcoming, sometimes the

improbability of the evidence offered by the defendant can be eyewateringly naive. This one then said that he had chopped up his mother-in-law's body and placed the pieces in rubbish bins around the city. We never did find any evidence of her, and there was certainly none to support the urban myth circulating in his home city about her postmortem presence in his kebab shop.

The second most common dismemberment classification is aggressive dismemberment, sometimes referred to as 'overkill'. This is a progression from a heightened state of rage, often reached during the homicide itself, which continues into the dismemberment phase, where it manifests in violent mutilation of the body. It is characterised by an almost haphazard, rather than logical, approach to the act. In such cases it is not unusual for the dismemberment to begin before the victim is dead and it can therefore sometimes ultimately be the cause of death. The analysis of patterns of injury assists in the determination of this category, which typifies the modus operandi of England's most famous serial killer, dubbed Jack the Ripper, who murdered at least five women, and possibly more than eleven, in the streets of Whitechapel in Victorian London.

Over a hundred possible identities have been suggested for Jack the Ripper. Disappointingly, there is little evidence to substantiate the claims made for William Bury, the last man to be hanged in Dundee, who was executed for the murder and mutilation of his wife Ellen and had lived in Bow, near Whitechapel, at the time of the killings. But if it was him, I have the neck vertebrae of Jack the Ripper sitting on a shelf in my office.

Offensive dismemberment, the third type, often follows a murder committed for sexual gratification, or results from the sadistic pleasure of inflicting pain on the living or meting out injury to the dead. This type of dismemberment frequently involves mutilation of the sexual regions of the body and may be the primary purpose of the murder. It is, thankfully, very rare.

Necromanic dismemberment, the rarest type of all, receives undue and disproportionate attention in films and novels because of its huge scope for the portrayal of gruesome and horrifying acts of violence and depravity. The motivation may be the acquisition of a body part as a trophy, symbol or fetish. Cannibalism also falls into this category. It should be noted that necromanic dismemberment is not always preceded by killing: it can take place, for example, when individuals have access to an already dead body, or involve the exhumation and desecration of corpses. In deference to humanity, decency and religious beliefs, we expect corporeal remains to be left in peace in perpetuity and while society can accept accidental disturbance or planned intervention for justifiable causes, what is not tolerated is abuse of the dead.

Finally, there is communication dismemberment, often used by violent gangs or warring factions as a threat to persuade their enemies to desist from a particular activity or to coerce others, generally young men, to join their gang and not a rival one. The message is powerful and clear: if you do not do what we want, this is what will happen to you.

In Kosovo, where I spent most of 1999 and 2000 as part of the British forensic team assisting the International Criminal Tribunal for the Former Yugoslavia, we saw examples of this form of 'communication'. A young man, usually an ethnic Albanian, would be abducted, murdered and dismembered into small pieces. Parts of his body would be left on the doorsteps of the families of other young men as a calling card advising them not to enlist in the paramilitary Kosovo Liberation Army. With some, it achieved the desired effect instantly. For others, though, it merely fuelled a nationalistic determination to join the guerrilla warfare against the Serbian militia.

◊

As designated experts in the UK, my team is often called upon closer to home to help with dismemberment cases. These can be complicated enough without the body parts being scattered across two different counties, as they were in one such case in 2009.

The first the police knew of this suspicious death was when a left leg and foot turned up, wrapped in plastic bags, by the side of a country road in Hertfordshire. It was very fresh and had been so cleanly detached from the torso at the hip joint that they thought it might be the result of a clinical amputation from a nearby hospital. They checked first with all the hospitals in the area to see if there were any irregularities with their waste-incineration procedures, but all were adamant it hadn't come from them. A search of the DNA database turned up no matches. It was clear that the limb belonged to an adult Caucasian male. Height could be calculated from its length, but on the basis of the limited information that could be gleaned from the leg alone, it did not seem to tally with descriptions of anyone listed locally as missing or with any individuals recorded by the Missing Persons Bureau.

Seven days later a left forearm, severed at the elbow and wrist, was found, again wrapped in plastic bags, in a ditch by the side of another road about twenty miles away from where the left leg had been dumped. The DNA was a match with the leg. Two days after this discovery, a horrified farmer in Leicestershire came upon a human head that had been tossed into his cow field. Because the report of the head was made to a different police force, a link with the previously matched leg and forearm was not made immediately. Leicestershire police had been looking for a high-profile missing person and wondered if perhaps the head could be hers—although the remains were very recent, facial identification was not possible because the skin and soft tissue were absent, probably, the pathologist believed, due to animal activity. But our analysis indicated that the individual was most likely to be male, and a superimposition of the skull on to a photograph of the missing person suggested that it was highly unlikely to be a match.

The Leicestershire police, too, made a fruitless search of the DNA database and for a few days two separate forces were independently hunting for missing body parts in their own areas. The following week, back in Hertfordshire, a right leg, cut into two pieces at the knee, wrapped in plastic bags and concealed in a holdall, was found in a layby on a rural road. Finally, four days later, the torso was discovered, with the right arm, from which the hand had been severed at the wrist, and the upper part of the left arm still attached, all wrapped in towels and stuffed into a suitcase which had been dumped near a field drain in the countryside, again in Hertfordshire.

A DNA link between all of the body parts was now made but, with no match on the national database, establishing the victim's identity, and therefore investigating his death and finding the person, or persons, responsible for it, was going to be challenging. Although the feet were still attached to the lower limbs, the hands had both been removed and were still missing, so the dismemberment did not quite follow the normal six-piece pattern. The distribution of body parts, however, was consistent with the most common motivation for dismemberment: ease of disposal. The absence of the hands and damage to the face pointed to the possible additional defensive intent of attempting to obscure the victim's identity.

Having a body spread over such a wide geographical area caused a bit of administrative hassle. Who should lead the investigation? The force that found the head? The force that found the first body part? The one in possession of most of the body? It is no small logistical problem to operate a major incident inquiry across different police forces, not least because of budgetary and staff implications. But as it turned out, this was one of the most professional collaborations we have encountered between any two police forces.

Dr Lucina Hackman and I headed south to assist. The long journey provided us with a lot of talking time—and if talking were an Olympic event we would take home gold every time. Although we made a little detour on the way to help with another matter involving a drug gang turf war and a facially disfigured body in the north of England, we spent a huge amount of time discussing the dismemberment case. We had a hypothesis that didn't coincide with the police theories and we needed those seven hours in the car to think it through and talk it out because, if we were wrong, we were going to look like the biggest couple of numpties this side of the River Tweed. But if we were right, there were going to be a lot of very hyperactive police officers running around Hertfordshire and Leicestershire.

We didn't agree with the suppositions relating to the MO of the dismemberment. There were a couple of things that just didn't fit for us and we are a suspicious couple of old biddies. Our first problem was with the sites of the dismemberment cuts. Yes, the pattern was almost classical, but the way it had been done was unusual. Those who have never dismembered a body before—and let's face it, that is the vast majority of the population—would be most likely to attempt to saw through the long bones of the limbs, the humerus in the arm and the femur in the thigh. Research at our centre has indicated that, when faced with the necessity of dismant-ling a body, most people would reach for a sharp kitchen knife first of all, and only when they found that, while it could cut through the soft tissue of skin and muscle, it could not cut through bone, would they head to the shed or the garage for a saw, usually either a wood saw or a hacksaw. Those accustomed to cooking might also consider using a hacking implement such as a meat cleaver from the kitchen or an axe from an outbuilding. But this body looked as if it had been 'jointed' rather than sawn into pieces, and that is very rare. In fact it was a first for us. We needed to see the surfaces of the bones to determine what type of tools had been used because there was definitely something odd going on.

Secondly, the head had been treated differently from the rest of the body. It had been found in a different county, for a start. It had not been wrapped up and it was the only body part where there had been a loss of soft tissue. We were not convinced by the pathologist's theory that this was due to animal activity as there was no sign of the normal scavenging pattern of tooth marks caused by the predations of either domestic or wild animals.

Tool-mark analysis is, in principle at least, very straightforward. If two objects come into contact, the harder of the two may leave a mark on the softer surface. If, for example, you cut a block of cheese with a serrated bread knife, then the knife, the harder object, will leave little ridges on the softer object, the cheese. The same thing happens with bone. If a bone comes into contact with a sharp object, such as a knife, a saw or the tooth of an animal, telltale marks are left on the bone and the pattern they form may be sufficient to identify the type of implement that made them. So if the loss of tissue on the head of the deceased man was the result of the scavenging activities of an animal, which we seriously doubted, we would expect to see characteristic marks made by its teeth.

The head had no skin and no muscle still attached. The eyes were missing, as were the tongue, the floor of the mouth and the ears. That would be an incredible feat for an animal to achieve without leaving tooth marks. What we believed we would find were cut marks, made by a sharpened blade, in the areas where the different muscles attached to the bone. If that proved to be the case, unless a common or garden badger had suddenly experienced an overnight evolutionary miracle and become dextrous with a knife, the tissue had to have been removed on purpose by a fellow human—and that would require some explaining. The head had been severed cleanly from the neck between the third and fourth cervical vertebrae, and there was something very unusual about that dismemberment site, too.

We kept our thoughts to ourselves until we could examine the head. At the briefing, we listened politely to the explanation of suspected animal activity. In such circumstances, both Lucina and I are aware of our eyebrows. We have very expressive eyebrows, we are told, and when we don't agree with something we are hearing, apparently they go up and down like uncoordinated hairy roller binds. We once found ourselves appearing as experts for the defence in an English court, where we were so incredulous at the evidence we were hearing from the Crown, and so conscious that we were at all times in full view of the jury, that our foreheads ached with the effort of constantly trying to countermand the involuntary movements of our eyebrows. We would both make lousy poker players.

We said nothing during the briefing, and did our level best to keep our eyebrows under control until we were in the mortuary, where we were able to look at the remains more closely. The head was remarkable for the skill with which all the tissues had been removed. We found knife marks exactly where we had anticipated we would find them, at the back and on the side of the head and under the jaw. All of the soft tissue was absent. In essence, the face had been completely stripped away.

The remarkable features did not stop there. Examining the other body parts, we saw that the dismemberment at the wrists had been accomplished by perfectly executed single cuts through the joint spaces between the carpal bones and the lower ends of the long bones of the forearm, the radius and the ulna. The hips were disarticulated, the femur having been removed from the acetabulum (hip socket), and the level of expertise demonstrated by the disarticulation around the humero-ulnar joint at the left elbow screamed at us that whoever had undertaken this task knew their anatomy well. What is more, they knew their human anatomy. And they had done this before.

It is extremely rare for a dismemberment to be performed without the assistance of a saw or cleaver of some kind but every part of this body showed clearly that no heavy or serrated implement had been used at any stage, just a sharp knife. Now, that takes real skill. There was no evidence of hacking or sawing even in the removal of the head. Indeed, it had been done in exactly the way anatomists, mortuary technicians or surgeons would do it to ensure minimum mess, fuss and effort. Forgive me if I don't share this secret.

Lucina and I went into a conspiratorial little huddle involving a lot of pointing and eyebrow animation. The police knew something was up and, sensing their agitation, once we were both utterly convinced, we called a meeting and broke the news. As always, initially they objected ('But the pathologist said . . .'). However, when faced with the irrefutable evidence, they took off, talking nineteen to the dozen into their mobile phones.

We speculated about possible occupations for the dismemberer. Vet? Butcher? Surgeon? Gamekeeper? Forensic pathologist? Anatomist? Surely not a fellow forensic anthropologist? Whoever this was, their tremendous skill at dismembering was not matched by an aptitude for disposing of the remains: everything bar the hands (which have never been found) had come to light pretty quickly.

The cause of death was straightforward: the victim had been stabbed twice in the back with a four-inch blade. One of the wounds punctured his lung and he may have survived for some time after the assault. The pathologist had estimated that the dismemberment could have taken up to twelve hours to complete but we disagreed. The level of skill suggested to us that the whole process could easily have been achieved in less than an hour, with maybe another hour or so required to pack up the body parts and clean up the premises.

Once our analysis is over, our photographs taken and our report written, we may never learn the outcome of an investigation unless we keep an eye on the newspapers. Because we work around the country, we don't always have the close relationship with the police that people expect from watching TV crime dramas and sometimes, as happened in this case, you hear nothing until you are served with a court citation several months later. We don't know what the police have found, we don't know the outcome of their investigations and so we go into court with only our evidence and often no context within which to place it.

I hate appearing in court. Operating in this alien arena is a stressful part of the job for scientists. Here we do not make the rules and we are rarely informed about strategy. In our adversarial legal system, one side wants to prove you are the world's leading expert and the other side wants to prove you are a blithering idiot. I have been cast as both and a lot in between.

In what had become known in the press as the 'jigsaw murder', once all the body parts had been examined, the police had been able to match the victim to a missing man from north London and dental records had confirmed the identification. His blood had been found in the bedroom and bathroom of his own flat and in the boot of his car, but there were only tiny spots. The killer and his accomplice—a man and a woman were facing trial—had cleaned up well.

The couple had been charged with the murder and other related theft and fraud charges. Two defendants meant three sets of lawyers' questions to deal with, plus a potential re-address from the prosecution. So four sets of questions to prepare for—deep joy! Testifying in an unfamiliar court in an unfamiliar city, on a case you worked on nearly a year before, has nothing to recommend it and every reason to make you nervous. If you have been called to give evidence, you assume there is a belief that you must have something of value to add to the proceedings but you have no idea what it is and in which direction the questions will be designed to lead you.

You are always questioned first by the counsel who has secured your services, in this instance the Crown. This is usually the gentlest of the interrogations but I always seem to stumble over the bit where they ask your age. It is not that I am in denial, but my age is just so unimportant to me that I often have to stop and think about it, which never fails to raise a titter around the room. My hesitancy lasts only a split second but it is enough to disarm me. Every time it happens I scold myself for forgetting to remind myself of my age beforehand but I never remember. In the circumstances, it is hardly uppermost in my mind.

The Crown took me through my qualifications and then my evidence, which was all quite plain sailing, but it lasted most of the morning and then the judge halted for lunch, which means you have to go away and kick your heels for an hour knowing that when you come back, you will be cross-examined by both sets of defence lawyers. This is where the adversarial element comes into play and where the challenges generally arise. It is perfectly possible for my stint in the witness box to run into a second day, which is even more nerve-wracking, especially as you can't talk to anyone about the case in between.

The first defence lawyer was charming, which is always a worrying start. Having accepted my qualifications, he wanted to talk about our assertion that the dismemberer had a detailed knowledge of anatomy. He informed me that his client was a personal trainer and a former nightclub bouncer. He had no anatomical training and had never worked in a butcher's shop, never lived in the country or pursued countryside activities. He certainly wasn't a surgeon or a vet, or indeed an anatomist or forensic anthropologist. How, then, could he possibly know how to make the incisions I said he had made or possess the skills that I alleged he must possess?

At times like these a cold sweat starts somewhere at the nape of your neck and begins to trickle down your spine. Could I really have got it so wrong? You question yourself again and again but you can't come to any other reasonable conclusion. The lawyer then turned to the matter of the implements. Surely, he reasoned, dismemberment would require specialist tools, to which I replied that a simple sharp kitchen knife would suffice for the method used in this instance, provided the perpetrator knew what they were doing.

'But surely the type of sharpness required would not be found in a domestic kitchen knife?'

I knew as the words fell out of my mouth that my response was likely to get me in trouble. 'With the greatest of respect, sir, the knives in my kitchen are sufficiently sharp to effect this.'

Quick as a flash he countered: 'Remind me not to come round to yours for dinner.' Laughter rang around the court and I was stunned. I had never experienced humour in a courtroom once a trial was underway, certainly not during a trial dealing with the desecration of a body as well as a murder. Perhaps I shouldn't have been so surprised. After all, death and humour have always been close companions and perhaps having endured such a gruelling few days, the court was grateful for a little levity to dispel some of the tension. I so wanted to retaliate with a witty one-liner but I did not dare. Trying to be smart is the quickest way to fall foul of a sharp-talking lawyer. So, very wisely, I thought, I buttoned it.

And then, suddenly, it was over. No cross from the second legal team and no redirection from the Crown. What I had anticipated would be the worst part of the process was over in the blink of an eye. It just goes to show that you can never predict what will happen in court, especially when you are not party to the legal tactics being followed.

Prior to, and throughout the duration of the trial, the accused and his accomplice had maintained their innocence, and then, without warning, towards the end they dramatically changed their plea to guilty. The man admitted to the murder and his girlfriend admitted to aiding, abetting and perverting the course of justice. The life sentence given to the killer was enhanced due to the aggravation of dismemberment, with very little reduction for the change of plea because it did not come until the case was all but concluded and because the crime was so severe. He received a minimum of thirty-six years.

Shortly before being sentenced, he confessed, through his somewhat shell-shocked lawyer, to being responsible for the dismemberment of at least four other men. This came as a total surprise to the police but he refused to elaborate on either the identities of his victims or the location of their bodies.

The convicted man had indeed been legally employed as a doorkeeper at a nightclub, but he was also a trained 'cutter' for a notorious London gang. If they killed an informant or someone else who was causing them some trouble, they would take the body round to the back door of the nightclub in the middle of the night. Here the cutter would dismember the body and pass on the pieces to the 'dumper', whose responsibility it was to get rid of them, often by burying them in Epping Forest, which was where our murderer said he had disposed of the deceased's hands.

He had been apprenticed to a senior cutter and learned on the job how to dismember a human in the most efficient manner possible. The fact that disposal of the body parts was handled by colleagues of a different grisly specialism explains why, while he was a truly skilled 'cutter', he was an absolutely rubbish 'dumper'. Who would have thought such activities could be someone's actual job? Imagine having those on your CV.

I had never been so relieved to find out that Lucina and I had been right. The reason for the removal of the victim's facial soft tissues was to conceal forensic evidence. Initially, the two co-defendants had each claimed the other had committed the murder, and by different means. Only examination of the face and the soft tissue of the head and neck could have confirmed which of them was telling the truth, so removing that was a form of insurance policy in the event that they were caught. They believed that if we couldn't prove who was lying, the court would be unable to come to a decision. So why they decided to change their pleas, we will never know.

Their motive seems to have been nothing more than financial gain. They stole their victim's identity in order to sell his possessions and clear out his bank account. He was an innocent man who gave two people a roof over their head when they were in need, and they repaid him by taking his life and desecrating his body.

◊

In court I never allow myself to be distracted by the actors in the room. The only people with whom I make eye contact are the barristers and the judge. I never, ever look at the accused. If I should happen to meet them in the street, I don't want to recognise them. I rarely look directly at the faces of the jurors, either, unless I am specifically asked to explain something to them, because I don't want to be diverted from the question being asked of me by their facial expressions. So I tend to focus on the shoulder of a juror sitting in the middle of the jury box. I don't let my eyes stray to the public gallery, where the anguish of families would unquestionably rattle my concentration. I marvel at their stoicism, though, especially in harrowing murder cases. What they have to hear is sometimes so personal and so brutal I can't help but wonder how they can bear listening to it being discussed in open court, with journalists writing down every distressing detail for publication immediately online and in the newspapers the next day. The family are victims, too, and their agony is often palpable.

The media have a remit to report death but the relish with

which they do so, and the often disrespectful headlines they use, can be distasteful. The more deviant the death, the more papers the story will sell. I am certain that such a predatory and exploitative mentality would not sit so comfortably with them if it were their own family that was being exposed to it, but as long as there are people who want to read about gruesome deaths there will always be insensitive journalism.

I am not sure I would be strong enough to deal with it if I were personally affected—certainly not if it were my daughter who had been murdered and my own son who was the perpetrator. This happened in one case in 2012 in which the glare of the spotlight was all the more fierce because the victim was an actress who had appeared in a television soap opera.

Gemma McCluskie was reported missing by her brother Tony the day after she was last seen alive. He made an appeal for her safe return and joined the search party out looking for her. All the time he was the very person who knew exactly where she was.

Gemma had been picked up on CCTV cameras as she returned to the house she shared with her brother in east London, which was also where the last call from her mobile phone had been made. Five days later, a suitcase was found in the Regent's Canal less than a mile away. It contained the dismembered torso of a young woman. Identification of a tattoo and subsequent DNA analysis confirmed that it was Gemma. A week later her legs and arms were found, wrapped in plastic bags, in the same stretch of water. It would be six months before her head was discovered, further upstream and also in a black plastic bag. It was only then that a cause of death could be assigned.

Gemma's brother, who was addicted to skunk cannabis, was quickly arrested for her murder. He was known to be unpredictable, sometimes violent, and she was reported to have been losing patience with his irresponsibility and escalating drug use. He confessed that they'd argued after he left a tap running and flooded the bathroom. He admitted to losing his temper but said he did not remember hitting her, killing her or dismembering her.

The cause of death was blunt-force trauma to the skull. All the elements of the killing and its dehumanising sequel were textbook examples of defensive dismemberment: a furious argument fuelled by drugs; assailant and victim known to each other; death occurring in the home of the victim; dismemberment undertaken by the assailant, unresearched and unplanned; the body being separated into the typical six pieces, wrapped in plastic, transported in holdalls and suitcases, disposed of in water close to the murder site and easily accessible. Early attempts at dismemberment failed and success was finally achieved with an alternative tool; both were implements that might be found in any domestic kitchen, a knife and a cleaver. As these characteristics pointed to an offender with no prior experience of committing such crimes, the strongest suspicion fell on Tony McCluskie.

As he continued to maintain that he had no recollection of what had happened, some of what follows is fact and some conjecture. What is certain is that Gemma had received at least one fatal blow to her head from a heavy object which was neither identified nor recovered. It is likely that she died where she fell. When McCluskie, probably high on drugs and enraged, realised that he had killed her, he panicked and, rather than standing up and taking responsibility for his actions, chose to try to hide them and plead ignorance.

It was a small house and there was nowhere her body could be concealed once the police came looking for her. So he decided that he had to get rid of her. He knew the only way to get her out of the house without being detected would be in parts. Where did he put her while he worked out what to do? We don't know. No evidence of blood was found in the bath, the likeliest site, and indeed a layer of undisturbed dust was noted here. Perhaps he laid her down on plastic sheeting on the floor, using towels to soak up her blood. Whatever he did, the surface on which he did it was protected.

At some point, he undressed her down to her underwear. Where on earth do you start to cut someone to pieces and how are you going to do it? And not just anyone, but your own sister? It would be a horrific prospect for anyone in their right mind and must have been driven by intense desperation. Perhaps he looked around the kitchen, saw the block of knives and decided to begin with one of those—certainly one of them was missing.

He began with the front of her right leg, using a serrated blade to cut across the limb about a third of the way between hip and knee. Needless to say this attempt failed, but he made around fifty-six cuts before conceding defeat. He then found something much heavier, perhaps a meat cleaver, and, having discovered that this was far more effective, stuck with it for the rest of the process. That he used the unsuccessful blade on only one part of her body told us that he was learning how to do this as he went along. In total there were at least ninety-five cuts, thirty-nine caused by the heavier implement on top of the fifty-six made with the fine-bladed knife.

Her torso was then squeezed into a wheeled suitcase. CCTV showed McCluskie lifting a very heavy bag into the back of a taxi. When the driver was tracked down he confirmed that he had been directed to the nearby canal and identified the accused as his fare. McCluskie probably returned afterwards with the limbs and head and dropped them into the water at around the same point, although there is no CCTV footage to verify this. Presumably he did not need to use taxis for those journeys as the body parts he had to carry were less bulky.

I received a summons to give evidence at the trial. I am not sure what it really added to the case being heard but my suspicion is that the Crown considered a full scientific account of the dismemberment and the number of cuts made to remove each of the body parts to be of value in showing how prolonged and callous the disposal had been. You choose your words very carefully in these circumstances, being only too aware that relatives are in the room. The last thing you want to do is to add to the monstrous pain and grief they have already experienced. We try not to be emotive in our language but there are few soft words you can use to describe such a heinous offence as criminal dismemberment.

I had to testify to precisely what was done to Gemma's body, confirming the order in which her limbs and head were removed and whether she was lying prone or supine for each cut. It was challenging to have to articulate all of this in front of her family and between the sobs and cries of pain it evoked. I was relieved that my evidence was accepted by the defence and that there was therefore no cross-examination, which spared the relatives having to hear even more details of what had happened to her.

I was in and out of the witness box within an hour, and about to leave court when the family liaison officer (FLO) stopped me and asked if I would be prepared to meet Gemma's father. He had spoken to, and personally thanked, everyone else who had been involved in his daughter's case and he wanted to meet me, too.

In our world, we strive to maintain a clinical detachment while engaged in our work and are largely removed from the immediacy of the grief and distress of family and friends. While I had met victims' relatives on overseas deployments, I had never before done so in the UK, and certainly none who had just sat through my evidence at a murder and criminal dismemberment trial listing the insults meted out to one of his children by another. I was very nervous and uncomfortable. What on earth do you say? What could you say? I could not, and would not wish to, experience his pain and I had no words that could in any way ease his family's agony. But he was not looking for anything from me. This was about him completing a task he felt it was his responsibility to fulfil.

As I waited in the witnesses' room for the officer to bring Mr McCluskie in, I was both hot and cold at the same time. The door opened quietly and a short, burly figure entered confidently. He was the kind of man you might expect to see propping up the bar of an East End pub; someone who in other circumstances would probably be the life and soul of a party. He shook my hand and sat down without a word. I could tell he was broken: there was a deadness about him and a well of sorrow behind his eyes. He was performing the last service he could for his lovely daughter: thanking the people who had enabled the truth to be told by playing their part in convicting his own son. With astounding fortitude, he had thanked everyone from the divers who took her body parts out of the canal to the SOCOs and the investigating officers, and now he was thanking the forensic anthropologist. In the face of the astonishing dignity, respect and sense of duty he displayed, my words were lame and redundant.

For as long as I live, the depth of that man's love for his daughter—and indeed for his son—will remain with me as a beacon of how humanity and compassion can triumph even amid the most appalling adversity.

CHAPTER 10

Kosovo

'More inhumanity has been done by man himself than any other of nature's causes'

Baron Samuel von Pufendorf, political philosopher (1632–94)

Day one in Kosovo.

Our world seems to grow smaller with every day that passes. Our constant craving for instant information on events taking place around the world has been fuelled by the rapidly advancing technology that can supply it. The days when our news was delivered by the papers every morning, and in bulletins on the radio or television broadcast at scheduled times, are long gone and what was once global now feels almost local.

It was cable television that first got us hooked on the twenty-four-hour-news habit. The ease with which a TV crew could transmit pictures around the entire planet from the site of an attack or disaster within minutes of it happening fed our demand for rapid information. In 2014, images of the smoking wreckage of the Malaysia Airlines plane shot down over the Ukraine were beamed around the world before the families of its passengers and crew even knew that a disaster had befallen their loved ones. I remember a time when news like that would be brought by a knock on the door, usually late at night, from a police officer, grave of face and with his hat tucked under his arm.

In the twenty-first century, even our round-the-clock news channels are no longer sufficient for us. The endlessly repeated cycles provide little new material with each telling, though we will try to squeeze every drop of information from what is on offer. Today, social media and mobile phones keep us up to date on the move, so we don't need to be in our sitting rooms monitoring the box in the corner to keep abreast of developments.

Of course, change is constant, and for the most part positive, and new technology has revolutionised our lives for the better, but occasionally I can't help thinking wistfully of a formidable Highland matriarch who, in days gone by, was horrified to learn that improvements to the mail system would mean she would have her post delivered every weekday. 'Is it not bad enough that I have to suffer bad news once a week?' she lamented. 'Now you want me to have it every day.'

Sometimes we forget that a simpler life has its benefits. So many of the news stories we follow are in truth of limited interest, and have no direct impact on our daily lives, yet we still want to know every last detail. We absorb most of it passively, even dispassionately, and I do fear that information fatigue is in danger of leaving us with the sense that the world holds little that can surprise us.

Death invariably has a starring role in the headline news, from her large-scale, though often impersonal, depredations in wars, famines and natural or humanitarian disasters to her seemingly random cull of individual loved and respected figures. She got a pretty bad press in 2016, when there was a collective feeling that she was taking more than her fair share of people well known to us all, although in fact there was no increase in the death rate that year compared with any other. Once such an idea takes root in our minds, we are inclined to view subsequent similar events as supporting such a misconceived theory—an example of a well-known forensic problem called confirmation bias, in other words, a tendency to seek out evidence that fits a pre-existing hypothesis.

In 2017, death seemed to be stalking the UK in the form of random terrorist attacks as the worldwide trend for using unsophisticated and simply planned and executed methods to injure and kill innocent members of the public gained ground. Mowing down pedestrians with a vehicle and then attacking them with common or garden domestic knives, as happened in London at Westminster and London Bridge, was first seen in the UK in the shocking murder of Fusilier Lee Rigby in 2013 and is a difficult form of attack for intelligence agencies to predict and therefore to prevent. Terrorism is by its very nature about generating fear. Our knee-jerk responses, such as placing security barriers on London's bridges, will of course serve some purpose, but those responsible will merely adjust their methods and adopt new ones. All we can do is remain intransigent to its tyranny and strive to keep one step ahead of the barbarism.

By and large, unless we have been directly affected by the events that make the news, the coverage death receives in our media fails to make a deep and lasting impression on our everyday lives. The war in a far-off land or activities of the despotic military regime that preoccupied our attention last week will inevitably fade from the news banners on our screens as we, the news consumers, move on to the latest celebrity revelation, reality TV scandal or political blunder. Until something, somewhere happens that changes our perspective. Suddenly, a story becomes very real and very personal and, before you know it, has come to dominate the direction of your life.

For me that moment came when I took a phone call one afternoon in June 1999 from Professor Peter Vanezis, at that time a Home Office pathologist at Glasgow University, where I was a consultant forensic anthropologist. I had known Peter for many years and so hearing from him was not an unusual occurrence. When he asked me what I was doing at the weekend, I told him, foolishly assuming he was going to suggest dinner, that I had nothing planned. 'Good,' he said. 'You are coming out to Kosovo.'

From that moment on, I was glued to the coverage of the crisis in Kosovo and hanging on the reporters' every word, trying to soak up all the information I could about a part of the world which, I am ashamed to admit, I had to go and look up on a map.

During the 1990s I had been aware, like everyone else, of

the atrocities unfolding in Bosnia and was shocked that, in this day and age, such things could be going on in a country right on Europe's doorstep. I was also aware that the news reaching us would be sanitised, as in such situations some stories will be deemed too distressing to broadcast. If what we do see and hear is disturbing, you can bet your bottom dollar there will be a whole lot worse happening on the ground. But it was still 'somewhere else', somewhere foreign, and someone else was taking care of it.

By today's standards, detailed and reliable information was slower to get out and it wasn't until more horrific images began to emerge that we all started to wake up to the true extent of the horrors being perpetrated on innocent people. Nothing on this scale of deprivation and dispossession had been reported in Europe since the Second World War.

Forensic anthropologists have little forewarning of when their assistance might be required in an international crisis, if it will be required, and if it is, how long they might be away. In a tip of the hat to the 1970s Martini advertising slogan—'Any time, any place, anywhere'—my team has been dubbed 'the Martini girls' (you probably have to be old enough to remember those rather cheesy television ads for that to mean anything).

As the crisis deepens, we try to construct a reservoir of background information, sourcing reliable journalism and launching extensive internet searches, just in case. Because we know that the only predictable thing about a mass fatality is that it cannot be predicted.

By 1998 it was becoming clear from intelligence coming out of Kosovo that the humanitarian situation was escalating to intolerable levels. The United Nations (UN) was in discussion with the Serbian president, Slobodan Milosevic, and his government, and working to secure the withdrawal of troops and militia gangs from Kosovo. The OSCE (Organisation for Security and Communication in Europe) was reporting humanitarian crimes on an unprecedented scale and armed attacks were allegedly being made on civilians—the elderly, women and children. While diplomatic and political negotiations can seem dreadfully dull and slow to the outside world, the process is fascinating when you start to see where, when, why and how events are happening and begin to identify your own very small but emerging place in the story.

Peacekeeping troops were sitting across Kosovo's borders, aware of the murders, rapes and torture taking place and desperate to receive the signal to enter. But nothing could be done until there was UN agreement that all peaceful attempts had failed irretrievably. The appropriate international protocols had to be observed and, while there were clear reasons for that, they seem to make no sense when you know that every single day of inaction will result in innocent people being butchered or driven from their homes. It should come as no surprise, then, when vigilante groups gain ground and fight back, or guerrilla warfare gathers momentum, as a population struggles simply to stay alive. It was a horrendously complex situation and not one that could be resolved quickly.

◊

The Balkan region is no stranger to conflict. It has been a hotbed of political and religious tension since 1389, when the infamous Battle of Kosovo, the vicious and bloody defeat of the mediaeval Serbian state by the Ottoman Empire, set Muslim against Christian for generations to come. It forged a mutual hatred and sense of injustice so deep that throughout the centuries it would erupt on a regular basis into brutal combat.

Emboldened by their success, the victorious Ottomans started to assimilate many of the Serbian Christian principalities, including Kosovo, which has remained a disputed territory ever since. From the middle of the twentieth century, an uneasy peace prevailed in the region as a result of the active repression of nationalism under the lengthy iron rule of Josip Tito, president of the Socialist Federal Republic of Yugoslavia.

But nationalistic fervour remained undiminished on both sides. That such intensity of emotion could persist so close to the surface in both communities so many centuries later is an indication of the almost genetic imprint of the hostility with which each group viewed the other. The strength of Serbian nationalism and the belief that Kosovo belonged to the Serbs by right is demonstrated by the inscription on the monument commemorating the Battle of Kosovo which, though the words may be attributed to the mediaeval Serbian leader Prince Lazar, were chosen for a memorial erected as late as 1953:

Whoever is a Serb and of Serb birth
And of Serb blood and heritage
And comes not to the Battle of Kosovo,
May he never have the progeny his heart desires!

Neither son nor daughter
May nothing grow that his hand sows!
Neither dark wine nor white wheat
And let him be cursed from all ages to all ages!

The Yugoslav constitution of 1974 gave Kosovo extensive autonomy and allowed it to be run largely by the majority Muslim Albanian population, the descendants of the Ottomans. The predominantly Christian Serbians bitterly resented this control over what they saw as their spiritual heartland and viewed the Muslim presence and power as an intolerable insult.

After Tito's death in 1980, others with different agendas were soon disrupting and challenging the fragile peace. In 1989, Slobodan Milosevic pushed through legislation that began an erosion of Kosovo's autonomy. The violent suppression of a demonstration in March of that year was the first obvious portent of what was to come and on the 600th anniversary of the Battle of Kosovo, Milosevic made reference to the possibility of 'armed battles' in Serbia's future national development. It would not be long before the Republic of Yugoslavia began to collapse.

Did each side have murder in mind from the outset, or did the barbarism simply escalate as the struggle intensified? Whatever the case, the Serb mission appeared to be to rid their spiritual homeland of the 'vermin' (I quote a word used to me) that had gained a foothold there. In short, genocide. No mercy was shown as the smouldering bitterness, nursed for over 600 years, was fanned into a slow burn that would eventually become a raging inferno.

The first significant trouble in Kosovo began in 1995 and the region erupted into armed conflict in 1998, partly as a consequence of the 1997 uprisings in Albania, which had put over 700,000 combat weapons into wider circulation. Many of them found their way into the hands of young Albanian men, seeding the self-styled Kosovo Liberation Army (KLA), which launched a major guerrilla offensive targeting Yugoslav authorities in the territory. Reinforcements from the regular forces were sent in to maintain order and Serb paramilitaries began a campaign of retribution against KLA and political sympathisers, resulting in the deaths of up to 2,000 Kosovars.

That March, a firefight occurred at the compound of a KLA leader in which sixty Albanians, eighteen of them women and ten children, were massacred by the Special Anti-Terrorism Unit of Serbia (SAJ). This brought widespread international condemnation and by the autumn, the UN Security Council was expressing grave concern about people being dispossessed of their homes by excessive force. As diplomatic efforts to ease the crisis continued, and with fears of the hardships that winter would bring for large numbers of displaced people without shelter, it took an activation order from NATO (the North Atlantic Treaty Organisation) for both a limited air strike and a phased air campaign over Kosovo to secure agreement to a ceasefire. It was agreed that Serbian military withdrawal would begin at the end of October but the operation was ineffective from the start and the truce lasted little longer than a month.

The first three months of 1999 saw bombings, ambushes and murders specifically targeting refugees trying to flee across the border into Albania. On 15 January, following reports that forty-five Kosovo Albanian farmers had been shot in cold blood in the village of Racak in central Kosovo, international observers were denied access to the area. The Racak massacre was a watershed moment for NATO. It mounted a campaign of air strikes which appeared to serve only to intensify the brutality being meted out to the Kosovar Albanians. The aerial bombardment continued, almost unabated, for nearly two months before Milosevic finally succumbed to international pressure and accepted the terms of an international peace plan.

Within days of the air operations being suspended, UN KFOR (Kosovo Force) peacekeeping troops moved into the region and Louise Arbour, the chief prosecutor for the International Criminal Tribunal for the former Yugoslavia (ICTY), requested that all NATO member countries be prepared to assist with the provision of gratis forensic teams. Suddenly, instead of passively watching these devastating scenes play out on the television news, I was going to be catapulted right into the middle of them.

When I took that phone call in June from Peter Vanezis, I could not have imagined the impact it was going to have on my life. At that point I had never worked as a forensic anthropologist outside the UK and I was terribly ignorant about the way this kind of operation would function in the field. I knew that there would be large numbers of bodies to be examined and identified, but I wasn't totally clear what my role would be, how I was going to get there, how long I would be away and what it all really meant. But given everything I know now, I would do it all again in a heartbeat.

It never occurred to me to say no. My husband Tom insisted that I had to do it, I had to go. He is an incredible man and I am blessed to have known him since we were friends at school. He was tremendous about the upheaval to family life. Beth was a teenager, Grace had just turned four and Anna was two and a half. We hired a nanny for the summer and I prepared for the experience of a lifetime without much idea at all of what that experience would involve. I certainly had no idea of the long-term repercussions it would have on so many of us.

Peter and other members of the British forensic team had been the first to enter Kosovo, on 19 June. I was to join them six days later. All I knew was that I would be flying from London into Skopje Airport in Macedonia, where someone would pick me up and take me somewhere to a hotel. The next day I would be met—somewhere else—by UN officials and escorted across the border into Kosovo, which was still strictly speaking a controlled military zone. I would then be staying somewhere in Kosovo for about six weeks. So that was the details sorted, then.

When I came out of the arrivals gate at Skopje Airport, I was completely unprepared for the overpowering heat, the noise and the sea of faces, all jostling for attention, looking for someone they knew or offering taxi services. As I hadn't a clue who was supposed to be meeting me or where we were going, it was more than a little nerve-wracking. I stood there staring across the forest of white cards being waved at arriving passengers in the hope that maybe I would see my name written on one of them, or at any rate something that seemed to be directed at me. It dawned on me with some alarm that I was in a foreign country where I didn't speak the language and my mobile phone didn't work. I had no idea what I would do if nobody turned up to claim me like a pathetic piece of lost luggage. If my mother had known she would have killed me. As it was, we didn't tell her where I was going until I had arrived, and by then there was nothing she could do about it, except cry and worry, which apparently she did for the whole six weeks.

Eventually I spotted a white card with a single English word scrawled on it in marker pen. It was at least a familiar one: 'Black.' In for a penny, I said to myself as I approached the man holding it and tried to engage him in conversation. Unfortunately, his English was as non-existent as my Macedonian, or indeed any southern Slavic language. French didn't work either and, given that my only other option was Scottish Gaelic, I knew I was scuppered. Unable to understand a syllable of anything the other was saying, we resorted to gestures. He motioned me to follow him and a lifetime's worth of well-heeded advice about not getting into cars with strange men ran through my mind. If my nerves had been in a state of heightened alert before, they were now in tatters and screaming at me that this was probably the most foolish thing I had ever done in my life. If I was murdered and robbed, or worse, somewhere on a quiet Macedonian road, I would have only myself to blame.

The man led me to a rust-bucket of a taxi with a roaring engine that puffed noxious fumes through the closed interior. He was probably keeping the windows shut against the pollution from the streets, but the air outside couldn't possibly have been any worse, especially after he lit his third cigarette. It felt as if I were being simultaneously cooked and gassed. We drove in silence for what seemed like miles, leaving the outskirts of the city behind and climbing into the mountains along dirt tracks that left plumes of dust in our wake. I was calculating how much damage I might do to myself if I jumped from the moving car (the presence of my passport in my hand baggage, which I was gripping tightly, was moderately reassuring—at least I would be able to take that with me) when we pulled round a bend and in front of us loomed a replica of the Bates Motel a decade after its heyday.

The windows were covered in dust and grime and there were slates missing from the roof. A mangy mutt was chained to a tree outside the front door, which was banging in the wind. Without a word, my driver, now firmly rooted in my mind as my prospective murderer, jumped out of the car, signalled to me to stay inside and disappeared into the building. It was now or never. I started to plan my escape, and how to retrieve my luggage from the boot, watching all the time for the driver to return.

As I put my hand on the door handle to make a run for it, there was a rap on my window and a loud squeal from someone, which I guess must have been me as I was the only person in the car. I cranked down the window and looked into the smiling faces of two strangers. In cut-glass Foreign and Commonwealth Office accents, they asked if I was, perhaps, Sue Black? They told me they were from the British Embassy and suggested that I might like to get into their car. They didn't think the hotel was at all suitable for me, and I have to say I agreed with them.

As the man went off to deal with my taxi driver and I busied myself with my bags, it occurred to me that this could be out of the frying pan, into the fire, except that now I was convinced I was starring in a James Bond movie rather than a Hammer horror film. I only had their word for it that they were who they said they were, and I still didn't know where I was going. But at least if they planned to kill me, they would be speaking to me in English while they did so. To my mind, that was an improvement.

Happily, they turned out not to be sadistic murderers but an utterly charming couple who did indeed take me to a very nice hotel in Skopje (right next to the airport where I had started my journey nearly four hours earlier). After a great meal in great company, I began to relax and slept like a baby that night, too exhausted to be scared any more. The following morning was spent on the inevitable paperwork in preparation for the time-consuming journey across the chaos that was the border, with its checkpoints, long queues of lorries going our way and equally long convoys of trucks trying to get out of Kosovo.

Having never been on such a deployment, I admit to feeling pretty edgy during the long drive. Border crossings were strictly militarised, entry and exit was by permission only and we knew that there were still snipers in the area, not to mention IEDs (improvised explosive devices) laid to welcome us. We crossed from Macedonia into Kosovo at the Elez Han gate and headed south-west across the most majestic mountain passes towards the city of Prizren.

Progress within Kosovo was slow and hazardous because of the condition of the roads—the potholes were bigger than craters on the moon. Drivers were armed and radio communications were tense: the Serbs had still not completely retreated and there were believed to be residual pockets of resistance. At one point, hurtling along far too fast for the difficult road conditions, we took a bend at high speed and the driver had to stand on the brakes as we found ourselves almost up the rear end of a tank. I think I might have squealed—again. I'd never realised I was capable of squealing like a girlie, but it seemed as if Kosovo was bringing that out in me. It may sound a stupid thing to say, but my goodness, tanks are simply huge, and very scary indeed up close. My heart was in my mouth until I noticed the red, white and blue colours of a tiny flag painted amid the green camouflage.

A flood of relief washed over me. It was 'one of ours'. As a proud Scot, never before had the Union Jack meant anything much to me as a symbol of my identity but I will never forget how I felt on seeing it that day, stencilled on to the fat end of a tank, in that inhospitable landscape. In that moment, when it really mattered, I acknowledged willingly and gratefully that it was the British flag that brought me a sense of protection, safety and belonging and calmed my growing fears.

There was no time to drop me off at our residence: it was straight to the first 'indictment site', where the rest of the team were waiting. At the end of the dusty road that marked our outer security perimeter, the first thing I saw was another huge tank, this time a German one. These soldiers were efficient, they were polite and they were putting their lives on the line so that we could do our work unmolested. At the cordon, and parked along the track, was a raft of vehicles: a few stoic and persistent journalists were moving with us like the camp followers of old. A site entrance and exit route had been marked out with crime-scene tape and our HQ was a white scene-of-crime tent set up further along the track, out of the range of prying cameras. It looked just like any other crime scene, and the familiarity of the set-up was oddly reassuring.

In the tent we donned our usual white crime-scene suits, double latex gloves and heavy-duty black wellington boots, sweltering in the 38° heat. Our policing support was supplied by the Metropolitan Police, and our security advisers were the anti-terrorist branch—SO13, as they were known at the time. Strange to think now that they were pretty quiet then, in the lull between Northern Ireland calming down and the rise of Al-Qaeda and so-called Islamic State terrorism.

The back story to this crime scene was sobering. On 25 March, the day after the NATO bombings began, a Serbian special police unit sacked the village of Velika Krusa, near Prizren, which is the second-largest city in Kosovo and the last major urban sprawl before the Albanian border. The villagers sought shelter in nearby woodland and could only watch as their homes were looted and burned. They had no alternative but to head for the Albanian border in a refugee convoy, even though they knew they were at risk of robbery, torture, rape and murder. Armed men halted the group, separated the men and boys from their families and herded them into an abandoned two-room outhouse. A gunman stood at the door of each room and sprayed it with automatic Kalashnikov fire. Their accomplices then threw straw in through the windows, soaked it with fuel and torched the building. It was claimed that over forty men and boys lost their lives that night. We don't know for certain what happened to the women and children from that group, but wherever they ended up, we suspect they did not survive, either.

Incredibly, there was one survivor, who would become a crucial witness in the international war crimes process, and for that reason the site was earmarked by the ICTY for forensic-evidence retrieval. The principal criterion for designating a scene as an indictment site was the existence of strong information, perhaps from a credible eyewitness, indicating the time and location of the incident, the demography of those involved and what had allegedly happened. The forensic team would be instructed to gather all relevant evidence, log it, analyse it and compile a report. If this corroborated the eyewitness's account, the incident would be prioritised in support of the charges of war crimes against Milosevic and his associates.

I was unaware at the time that it was Peter Vanezis's arrival at Velika Krusa that had prompted his phone call to me. Surveying the scene in front of him, he had apparently said, rather graciously, 'I can't do this, but I know someone who can.' No pressure, then.

There is nothing glamorous about working in a white scene-of-crime suit, black rubber police wellies three sizes too big, a face mask and double latex gloves in searing heat. Thus attired, I stood at the door of the charred shell of the outhouse and looked in on a nightmare scene that could never be adequately described. The central door of the building led into a short corridor with one room to either side. There were at least thirty bodies in one room and another dozen or more in the other, all piled on top of each other in the corner diagonally opposite the internal doors, all badly burned, all extensively decomposed and all buried under fallen roof tiles.

They had been there for three months as the Kosovo summer heated up, readily accessible to insects, rodents and packs of wild dogs. They were boiling with maggots, fragmented and partly scattered and eaten by the scavenging animals. There was only one way to clear the space and that was to strap on knee-protectors, get down on your hands and knees and work systematically inwards from the door, lifting and sifting every piece of debris down to floor level. As well as retrieving all body parts and personal effects such as clothing, identity papers, jewellery or other items that might be identifiable by family and friends, it was vital that we collected all evidence relating to the crime, which included bullets and casings, as it might be possible to link them at a later date to a specific weapon and from there to the person who fired it, their commanding officers, and so on all the way to the top. This is a 'chain of evidence'—and as we all know, a chain is only ever as strong as its weakest link. We did not want that to turn out to be the forensic evidence gathered by our team.

You can't wear thick rubber gloves for work like this because you need to be able to feel what you can't necessarily see. Bone feels like bone, and really nothing else, and it was necessary to begin to process a body part as soon as we came upon it. We would clear around the approximate corporeal shape of an individual to try to isolate one person at a time, although this was challenging given the commingled nature of the scene. The heat was ferocious, the smell almost unbearable and the constant drip of sweat down your back, into your gloves and off your forehead into your eyes, which left them constantly stinging, was unpleasant in the extreme.

We were warned to be on the alert for IEDs, which had been found in such sites in the past—indeed, a device had already been discovered just before I arrived, connected to a trip wire across the path and designed to maim rather than kill. I had never seen a bomb in my life and wouldn't have recognised one if I found it in my porridge. I related my concerns to our SO13 explosives expert, who was an absolute gem. He said the best thing I could do if I came across anything at all that worried me was just stand up, call it in and leave the space. They would then suit up and check it out for us. He also advised me not to delve into the pockets of clothing as there had been reports of razor blades and hypodermics being placed there, again with the aim of causing injury rather than killing. He looked me in the eye and said, very slowly and clearly, 'Whatever you do, never, ever cut a blue wire.' Talk about messing with my mind. As if I was going to cut anything: I was far too bloomin' terrified.

Picture the scene: me, sweat dripping down my face and down my arms into my latex gloves, on my hands and knees sifting through rubble, face to face with boiling masses of maggots and rotting tissue, when suddenly, I spot the glint of metal. How much bravery am I going to show? Not a shred—the full double-yellow streaks run right down my back from top to bottom. I called it, we retreated and the explosives guys suited up and went in. They seemed to be in there for hours. When they walked back up to the base, where we were all standing around kicking dust, their faces were grave. They stripped off their body armour and the lead officer came over to me. He stood very close and his mouth was almost touching my ear as he told me, very clearly, without a hint of paternal compassion, 'You will never understand just how lucky you are to still be alive, little lady.' As he raised his hand into my eyeline, I could see that he was holding a shiny soup spoon.

Well, how was I supposed to know? Needless to say, I was

ribbed mercilessly for days by my team-mates. If a bowl of soup was delivered at mealtimes, mine would come with four spoons. I found spoons in my kit bag and even in my bed. I became the cutlery queen of Kosovo. I bore the jibes with good humour because they were a sign I'd been accepted into the fold. These were good, kind men and if they went to the bother of teasing you, it showed they liked you.

At that time I was the only woman on the team, which in certain circumstances might have proved tricky for some, but it was no problem for me. As a mother of three, it was easy and natural to adopt a maternal role. I listened to everybody's woes, sent them off to bed when they'd had too much to drink, offered advice and was generally non-threatening. Everyone was given a nickname—John Bunn was 'Sticky', Paul Sloper was 'Slippy'—and I would have been happy with Mother Hen or some such endearment. Unfortunately, however, I managed to earn myself a rather racier soubriquet by opening my own big mouth. Well, there's a surprise.

After we had cleared the first room and were about to start on the second, a press day was arranged. A gaggle of foreign ministers, including our own secretary of state for foreign affairs, Robin Cook, would be descending on us to see for themselves what conditions were like on the ground. Mr Cook and his entourage arrived by helicopter and gamely donned their white suits to come down to the burned-out building. I had decided I was going to dislike him purely on the basis that he was a politician. I certainly never expected to find myself truly warming to him, let alone admiring him. He did the required grandstanding for the cameras but when they were turned off, his microphone was removed and he stood at the door beside me, looking into that second room, he was visibly shaken by what he saw, no doubt imagining the horror suffered by those men and boys just a few months earlier. He said to me, 'If I close my eyes, I can hear their screams and I can feel their pain. How could this be allowed to happen?' He was doing exactly what we cannot allow ourselves to do: he was living it, and I respected his humanity and honesty.

As we came out of the house and walked up the dust track towards our decontamination station, all we could see behind the cordon running alongside it were rows and rows of camera crews. Every long lens at the site was trained on our party. I turned to my SIO, one of the most senior officers in the Met, and uttered a remark that provided him with a nickname for me which he uses to this day. As I peeled off my crime suit I quipped that, being the only woman in the team, I must look to the camera crews like the camp whore. From then on, in every Christmas card and every phone call, he would greet me as CW. It horrified my husband. But it was this kind of nonsense that kept us all going, even at the most sombre moments. As so often when we are in the presence of death, gallows humour dispels the tension. And it could have been worse. One of our pathologists, who arrived in a later team, and who shall remain nameless, was secretly referred to as 'Dagenham' because she was two stops short of Barking.

We cleared both rooms of the house at Velika Krusa and assigned as much of a biological identity to each body as we could, recording any individuating characteristics. Where we could establish a cause of death, it corroborated the eyewitness testimony, as gunshot injuries predominated. The oldest victim was probably in his eighties and the youngest around fifteen—not, in the eyes of his killers, a boy but a man to be wiped off the face of the earth before he took up arms against them.

Each body bag was given a number, all the personal effects were collected and samples of bone were taken for DNA analysis. Confirmation of personal identity would not be swift, not only because of the level of decomposition and the fire, but also because the Serbian forces had stripped many of the victims of their identity documents. We retained the personal effects and clothing and cleaned them so that they could be viewed by the families of the missing as a tentative means of identification. A preliminary determination of identity would need to be confirmed through DNA but in the meantime, the body would be assigned a URN (unique reference number) and released to the families for burial.

We had one mortuary tent equipped with a stainless-steel table for the postmortem examinations, but the initial triage of remains was carried out in the courtyard of the blackened outhouse, where we processed the evidence. We balanced two long planks of wood between the lip of a well and the back of a tractor to serve as a table. There was no electricity, no running water, no lights, no toilets and no rest areas. Our work in the field is rough and ready but ingenious. If I could choose, I would far rather be doing my job in an environment where we are challenged by real logistical difficulties than trying to do it in more comfortable conditions hampered by impenetrable red tape. At the forefront of our minds the whole time was that the quality of evidence collection was paramount. And I'm proud to say that forensic evidence from the British forensic team was never once questioned in the International Criminal Tribunal.

Although quality of evidence was our prime driver, it was equally important to us to maintain the dignity of the deceased and respect the grief of the living. This principle came to the fore when we took over a disused grain store at Xerxe, northwest of Prizren, as a temporary mortuary. In the early stages, there had been few onlookers interested in our activities but as the refugees began to return from Albania, the privacy we had been afforded to do our work could not be sustained. So it made more sense for the team to be split into a recovery team, which would bring in the bodies, and a mortuary team operating securely from a closed building, rather than to have both teams working together at the crime scene.

We had just taken possession of a fluoroscope, which gave us X-ray capability, and we now had the luxury of a roof over our heads, running water from a garden hosepipe and electricity from a temperamental generator, the noisiest contraption on the face of the planet.

The bodies were queued up waiting for postmortems and examining them was almost like working on a factory production line. We had a deadline, too, as a mass community burial was due to be held. We worked day and night to finish by the Saturday chosen for the funeral. This was the first ceremony of its kind to take place in Kosovo, and although we knew it would turn into another media circus, we were not prepared for the colossal invasion of camera crews that turned up at our little mortuary and camped outside in the car park overnight. They were desperate for pictures and comment and when nothing was forthcoming, tempers outside in the unforgiving heat began to rise. As it was thought they might be kinder to a woman, I was sent out as the sacrificial lamb to talk to them and give them what I hoped would be enough to defuse their mounting frustration.

The bodies were due to be picked up for the funeral from our mortuary by the families. Most of them would bring little open-top trailers towed by a tractor or by something resembling a ride-on lawnmower. The procession would then move up the hill to the burial ground at Bela Cervka. With so many bodies, it was going to be a long, long day. Security at Xerxe was provided by the Dutch military, who were camped not far away in a disused winery at Rahovec, and we were so concerned about the presence of the media that they supplied reinforcements to guard the mortuary overnight, helped by local volunteers. Before the first family arrived, I gave some interviews and was taken aback, not to say a little frightened, by the ferocious onslaught of questions and the aggression directed at me and our team.

At one point a reporter shouted: 'Are there children in there?'

'Yes,' I replied politely. He then asked if I knew where they were in the mortuary. Again I replied in the affirmative. He then demanded to be shown the bodies. I courteously but firmly declined his request. At that point, he very publicly called my parentage into question and invited me to perform a sexual act on myself. To say that I lost any scrap of sympathy I had for them at that moment would a huge understatement. I was determined that we would continue to protect the dignity of these remains and anybody who thought differently was in for a very long wait.

In making sure that imperative was met, I suppose I crossed a line from the professional to the personal. Maybe it was wrong, but I would do it again. There was no way they were going to get any images of these children leaving the mortuary if we had anything to do with it. Through our network of locals, we got a message out to the relatives who were due to collect children for the funeral, explained the situation and asked if they would be prepared to delay until later in the day. They readily agreed. It meant that we were not releasing the childrens' remains until well into the afternoon, by which time a large crowd was already gathering some distance away at the cemetery. This presented the media with a dilemma: if they hung around at the mortuary in the hope of catching a small coffin being carried by grieving parents, they risked missing most of the mass burial at the cemetery. Any who took the gamble went unrewarded. The children left in unremarkable adult coffins, but nobody except the families knew that, and were the very last bodies to leave our mortuary. There are many pictures out there of that day at the cemetery, but none that even hints that any individual victim was a child. A small pyrrhic victory perhaps, but it was one that really mattered.

The families were so appreciative that they honoured us by asking us to join the funeral procession. It was incredibly moving to walk behind the last trailer and to be gathered into the embrace of their collective grief. As we walked along, the women offered us tea and cool water. The tea we could take because the water had been boiled, but the water we just had to pretend to drink. So many wells in the area had been spiked with the dead that contamination was rife. We were a small team and we could not afford to have anyone succumb to illness, but at the same time we were desperate not to offend. They were offering us the only gift they had to give.

We came to witness these mass funerals too many times over the next two years as we worked through the indictment sites in Kosovo, but none was ever as personal as the first at Bela Cervka.

I made two more tours in Kosovo in 1999, each lasting six to eight weeks, and another four the following year. I was honoured to be a part of the first international gratis team to enter Kosovo, and just as proud to be a member of the last one to leave. We worked twelve- to sixteen-hour shifts, often seven days a week, and by the end of a six-week stint you were ready to go home—if you weren't, it was a clear signal that you needed to go home.

It can be a strange and almost enticing experience to be so cut off from the rest of the world, and for some, those who were perhaps not happy in their job or personal life, it was an escape. We had little idea of what was going on anywhere else, who had died, the latest box-office release or the next development in a juicy scandal. By the end of a tour I couldn't wait to get home to see my family and have a dose of normality.

We had intermittent access to a satellite phone, which kept us sane by allowing us to touch base just often enough with those we cared about. I remember feeling very homesick one night and calling Tom, lamenting how far away I felt from him and the girls. He asked me what kind of a night it was and I said it was glorious, the sky was clear and the moon so bright.

He told me he was sitting outside our house in Stonehaven, on the garden bench, looking up into the sky at the same moon. So we weren't actually that far apart after all, were we? I love a full moon and I really love my husband.

Every situation we experienced was different. Although there were general protocols to follow, every day brought some new challenge and some unexpected event. While we now had our mortuary with a roof, not every postmortem could be undertaken there. Often we would find ourselves tramping through countryside to reach remote crime scenes that were inaccessible to our vehicles. If we couldn't transport the bodies to the mortuary, the mortuary had to go to the bodies and we would be literally working in the field.

One day we were led to an extremely isolated site, about an hour's walk across rough terrain to an open patch of grassland up in the hills. Here, it had been reported, the elderly, women and children had been separated from the men in their refugee convoy, who had been moved on elsewhere to their own fate. The children had been taken to the other side of the meadow and told to run back to their mothers. They did so eagerly because they were so scared. As they crossed the open patch of grassland, with their mothers and grandparents forced to look on in horror, the captors took pot shots at the children. Once all the children were dead, they turned their guns on the women and the old people.

I don't know how even to begin to articulate the cruelty, the inhumanity, the torture of such a cold and calculated murder of innocents. We knew this was going to be hard for everyone and as we hiked closer and closer to the grave site, our mood grew increasingly sombre. Sometimes you welcome the odd flash of humour to lighten the atmosphere, but there were no attempts at levity that day. This was a despicable place where unspeakable acts had been perpetrated for the sport of barbaric men.

We spread out some plastic sheeting on the ground and the bodies were exhumed one by one from their mass grave. Remains that have been buried are likely to be better preserved for two reasons: the temperature below ground is cooler, reducing insect activity and slowing down decomposition, and they will have been protected from predators. But sometimes their good condition makes them harder to deal with from another perspective. The bodies are going to be more recognisable, and that can make it more difficult for the members of the team to reach the dispassionate place their minds need to occupy.

At one point a two-year-old girl was laid out on the plastic in front of me, still dressed in her sleep suit and red wellingtons. My job was to undress her, to let the police officers seize the clothing for evidential purposes and then to begin the anatomical survey of her body, cataloguing the ballistic injuries that had so devastated her tiny form.

Suddenly, I sensed a change in atmosphere. We had all been very quiet that day anyway but a new, heavier blanket of silence had descended. I looked up and all I could see in front of me was a long line of black police wellies and white crime-scene suits. For a moment, I was puzzled as to why everybody was standing in a row blocking me from view. It was only when I got to my feet that I realised what they were doing. One of our team had made the cardinal error of mentally transposing the face of his own young daughter on to the mutilated body of this little girl and he was finding it difficult to cope. The only way my male colleagues knew how to help him was to shield him from the sight of the dead child while he attempted to compose himself.

The mother on a team can't allow that to be the way to handle such a situation. So, without a word, I took off my gloves, rolled my suit down to my waist, walked behind the cordon of men and threw my arms around him until he had finished sobbing his heart out. I think it dawned on the men on our team that day that they didn't always have to be tough. Sometimes, especially when it comes to the appalling death of an innocent, someone has to shed tears, and there was no reason why it shouldn't be one of us. Having chinks in your armour isn't always a sign of weakness. It is often a sign of humanity.

At the end of our last, and very long, tour in 2000, the police sent out a team of counsellors. We had been in Kosovo for eight weeks solid by then. Living cheek-by-jowl with your colleagues for that length of time, you get to know each other incredibly well and the team becomes a second family. Forged into a close unit by our common purpose and experience, we supported each other when the need arose, and the intervention of outsiders, though well intended, was not welcome.

The counsellors gathered us all together in a nondescript room and sat us in a circle. They asked us to wear name badges so that they could create a sense of intimacy. We all knew each other's names, so it was solely for their benefit and we resented that. They were the ones who didn't know who we were, and neither could they ever comprehend our shared experience. We had lived with each other, fought with each other and cried with each other; we had drunk together and worked ourselves into exhaustion. But we tried to do our duty—or at least, most of us did—obediently sitting round in a circle while our badges were written out for us and stuck on to our chests.

The counsellors asked us how we 'felt'. How the bloomin' heck did they think we felt? We were tired and we wanted to go home. We had just spent two months up to our elbows in the detritus of a war that had indiscriminately killed men, women and children and we didn't take kindly to outsiders poking around inside our heads and rattling those cages.

Our mortuary technician, Steve, an outspoken Glaswegian, was the focus of specific attention. While the rest of us were sporting badges bearing our first names, his read 'Alf', which gave rise to some barely contained hilarity. Steve had been the tour prankster, and most of us had been the victims of his penchant for practical jokes. One of them had involved hiding a bright pink plastic novelty alarm clock—it was in the shape of a mosque and, instead of the usual beeping or ringing alarm tone, featured a *muezzin* calling the faithful to prayer—under the bed of one of the police officers, set to go off at full volume at 4am. As the *muezzin* got into his stride, Mick had shot out of bed, tripping over his boots, and vowed revenge. This, it seemed, was it, because it was Mick who was in charge of the black pen and labels. Why Alf? It stood for 'Annoying Little F**ker'. The mayhem that ensued every time the poor counsellor asked, 'So, Alf, how does that make you feel?' was simply exquisite and a much-deserved payback. Needless to say, the counsellors lost any hope of controlling this feral team.

These were the moments that counterbalanced the daily horrors, and they are the ones that stick with you. Moments shared in the private language of a camaraderie that only the people who were there with you can ever understand. It was a sobering time, but a precious one, and an experience I would not trade for the world. It tested me by teaching me how deep my abilities run, so that when I need to draw on them now, I know how far I can dig. In the process, I made friendships that have lasted for over twenty years. And irrespective of the passage of time, the unwritten team rule applies: when a Kosovo colleague needs you, you respond.

You cannot help but be indelibly marked by world-changing events such as the Balkan wars when they become your personal experience. You might come to count your own blessings more appreciatively, you might take up the cause and become politicised, you might immerse yourself fully in a new culture. Whatever you do, one thing is certain: you will never again be quite the same person as you were before. There are many things I would like to have changed about that time but none that I would swap. I learned a great deal about life, death, my profession and myself as a person. And one other vital lesson that will always stand me in good stead: never, ever cut the blue wire.

CHAPTER 11

When disaster strikes

'Show me the manner in which a nation cares for its dead and I will measure with mathematical exactness the tender mercies of its people, their respect for the laws of the land, and their loyalty to high ideals'

Attributed to William E. Gladstone,
prime minister of the UK (1809–98)

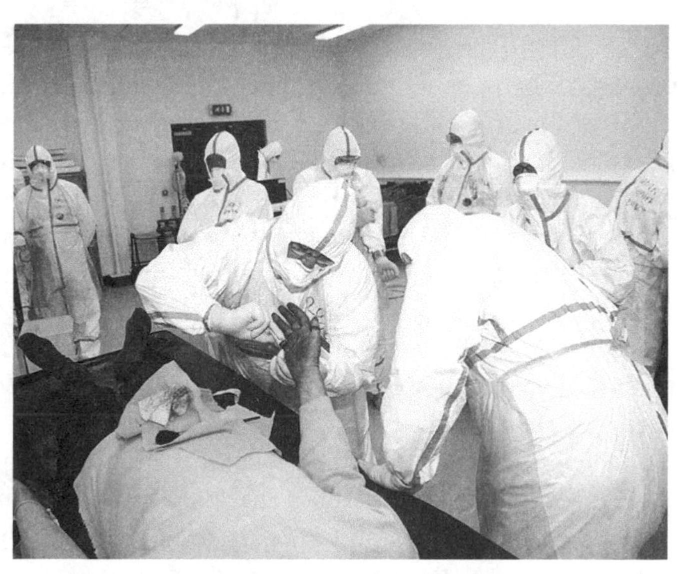

Fingerprinting in Disaster Victim Identification training.

On Boxing Day 2004 people across the world looked on, horrified, as a tsunami ripped around the Indian Ocean coastlines of Thailand, Indonesia, Sri Lanka and India. Most of us had had little occasion to use the word 'tsunami' (which means 'harbour wave') before that day, but it would be peppering everybody's conversation for months in the devastating aftermath of one of the worst natural disasters in recorded history.

No country in the world is immune to mass fatalities caused by disasters, whether these are brought about by a natural catastrophe or occur accidentally, as a result of human error or corporate negligence, or are deliberately perpetrated by acts of terrorism. For the sake of decency, health and justice, the deceased need to be managed, and this is best achieved through the well-rehearsed process of Disaster Victim Identification (DVI). To work successfully, DVI requires preparedness, advanced communication networks, inter-agency co-operation, crisis-management capabilities, efficient implementation of emergency plans and a swift response from trained personnel. It is complex, challenging, time-consuming and it is expensive. The incontrovertible fact that a mass-fatality event cannot be dealt with quickly, cheaply or easily needs to be accepted by every government around the world unfortunate enough to have one happen on its soil. History has shown us that if we do not pay attention to our dead in a proper and dignified manner, someone will be held to account and this can, and has, resulted in the fall of governments. It is a very serious business.

A mass-fatality event is often defined as an emergency situation where dealing with the numbers of injured and deceased, or parts of the deceased, exceeds the capability of a local response. This comfortably flexible definition, which does not attempt to quantify the casualties, reflects the fact that some regions have greater resources at their disposal than others and are able to cope at the same time as meeting the everyday demands of their region. The UK has seen its share of mass-fatality events and while, mercifully, the numbers have often been low enough for most areas to manage, there have been several major incidents in living memory in which more than a hundred people have lost their lives. Among them are the Gresford mining disaster of 1934, when an explosion killed 266 men and boys in north-east Wales; the still unexplained fuel explosion on board *HMS Dasher* in 1943, which claimed 379 lives in the Firth of Clyde, and the infamous Aberfan colliery disaster of 1966, when a wall of coal slurry crashed through a junior school. A hundred and sixteen of the 144 who died there were children. In 1988, Scotland was the scene of two major disasters: the Piper Alpha North Sea oil and gas explosion killed 167 men, and 270 lives were lost when Pan Am flight 103 was blown up over Lockerbie by a terrorist bomb concealed in a suitcase in the hold.

Given the multinational nature of our modern world, it is inconceivable nowadays that any event large enough to cause a mass fatality anywhere will not involve citizens of other countries. The horrific fire in 2017 at Grenfell Tower is a case in point. This requires us to think globally and to work in international partnership. The country where the event has occurred generally takes the lead, and it is incumbent on responders from other nations to be cognisant of relevant customs and laws. Forensic experts are always desperate to roll up their sleeves and just get on with the job—a job that nobody else in their right minds would ever want to do—but while the need to cross borders to recover and repatriate fellow citizens who have died on foreign soil may well be urgent, there will be diplomatic, governmental and legal hurdles that must be overcome first. As I found when working in Kosovo, in some countries you may not be able to deploy as swiftly as you would like, and this can be incredibly frustrating.

The most devastating tsunami in history was triggered by an undersea earthquake off the coast of Sumatra—the second-largest seismic event ever recorded. The massive oceanic waves it generated wreaked death and destruction in fourteen countries bordering the Indian Ocean. To say that they were caught unprepared would be an understatement. As this sort of natural disaster is extremely rare in the Indian Ocean, there were no early-warning systems of the kind introduced around the Pacific Ocean, where the possibility of volcanic eruptions and tsunamis was considered more likely. Over 250,000 people died, a further 40,000 were reported missing and millions were displaced. Over half the fatalities were in Indonesia, and the highest proportion of European casualties were among those on holiday in Thailand at the height of the winter tourist season.

As Tom and I watched the first reports coming out of Thailand on our television screen that Boxing Day, he looked over at me and said, 'You may as well pack your bags—you know you will be going.' As it turned out, I knew no such thing. Across the UK, forensic practitioners waited and waited for the call that would take them and their skills out to the devastated areas, all straining at the leash to jump on a plane. The silence from government was deafening. Eventually a tiny little press release was issued, announcing, with no fanfare whatsoever, that the Metropolitan Police were sending out some fingerprint officers. *What*?

It was the last straw as far as I was concerned. I sat down and wrote one of my 'middle-aged, red-headed, irascible Celtic woman' letters to the prime minister, Tony Blair, reminding him that forensic scientists and police had been quite clear on many occasions about how crucial it was to set up a DVI response capability—on the basis not that it 'might' be needed, but that it would be needed. It was vital that the UK was ready to react the moment a disaster occurred. Yet here we were sending out a measly handful of police officers when what the situation demanded was a strong DVI presence like the one provided four years earlier in Kosovo. We were nothing short of a global embarrassment, in my humble opinion.

I told Mr Blair that I considered national capability preparedness a matter of such urgency that, if a prompt response from the government was not forthcoming, I would be writing to express my views to the leaders of the other two main political parties. The silence just grew louder. I kept my word and sent copies of my letter to Michael Howard and Paddy Ashdown, then at the helm of the Conservatives and Liberal Democrats respectively. Perhaps inevitably, someone leaked my letter to the press and all hell was let loose—just as I was heading out of the country. Tired of hanging around for a call from the government, I had accepted the invitation of Kenyon International, a private DVI company, to fly out to Thailand, leaving poor Tom to field the calls from the press and broadcasters.

I saw in the New Year somewhere above Switzerland. With virtually every passenger bound for Bangkok for reasons related to the disaster, many of them on their way to try to find missing family members, it was an eerie and melancholy affair. As 2005 dawned, the pilot, who was himself well aware of why the flight was so full, came on to the tannoy and said that, while in normal circumstances we would all celebrate with some bubbles, this New Year he would ask us simply to raise our glasses in silence to those who had lost their lives, those who lost their loved ones and those who were on their way to help. It was a very emotional moment for everyone.

Thailand was in total chaos. Press and frantic families were flowing into the affected areas and the local services were struggling to cope. The depredations of the tsunami seemed haphazard. Vast swathes of land were laid to waste while others had been left miraculously unaltered. You would see a hotel standing apparently unscathed amid the rubble of all the buildings that had once surrounded it. Usually when working in a disaster zone, we expect to be sleeping on military cots or perhaps even under canvas, so it was somewhat disconcerting to be spending our days in the deprivation and despair of the makeshift mortuaries and then returning at night to a luxury hotel and its restaurants, bars and swimming pools. It just didn't feel right. When I was invited to use the hotel's laundry service, and questioned what seemed like an obscene luxury, I was reminded that the economy still badly needed to generate income from its paralysed tourism sector. We were the closest thing to tourists that Thailand was going to see for a wee while.

I had just finished unpacking my suitcase when my phone rang. It was my old senior Met officer friend from Kosovo. 'CW, are you still causing trouble?' he asked. 'I do hope so.' I could hear the amusement in his voice as he told me he had been instructed to liaise with me officially to find out whether I had a genuine concern over the need for a UK DVI response force or was just stirring up discontent. Oddly, he didn't wait for me to answer. Instead he warned me to expect a phone call within a few minutes from the prime minister's private secretary.

Sure enough, the phone rang again, this time with an invitation to a closed meeting to discuss a DVI response capability for the UK. The Downing Street official assured me courteously that this could be arranged to suit my diary. My goodness me. It had never been my intention to give the government a headache, less still a bloody nose, and I was well aware that if I wasn't careful it could rebound on me. I had spoken out simply to voice my firm belief that it is the duty of all those who have the skills required in a mass-fatality event to provide national assistance when the need arises. In a world where accidents, forces of nature and malign intent can strip away the lives of those we care about in the blink of an eye, we need to be standing by, fully trained, to handle the aftermath swiftly and professionally. Differences and egos have to be put to one side and everyone must pull together for the common good.

Good grief, but the heat and humidity in Thailand were overwhelming. One of these days I will be sent to a cold country more climatically suited to a redhead. Graham Walker, who was to become the UK's first DVI commander, once remarked to me on the craziness of a job where we think it perfectly normal for one of our colleagues to pass out through heat and exhaustion. And what do we do when that happens? We just prop them up against a wall, give them some water and, when they feel better, consider it quite OK for them to go back to work. Unusual people, unusual circumstances, extraordinary commitment.

The biggest enemy of victim identification in hot and humid countries is rapid decomposition. So a fast response and attention to the preservation of remains are a priority. A record of the locations where bodies have been found is also extremely helpful in speeding up the identification process, especially when people have died somewhere you would expect them to have been at the time, such as in their homes or on the premises of the hotel where they are registered. Thailand was to make our job even harder by confounding us on both counts.

The bodies from each town were being taken to local temples. When we arrived at our first collection site, Khao Lak, the scene we encountered outside the temple was truly grim. In an attempt to help, people who had access to vehicles had been collecting corpses all round the country and delivering them en masse to the entrances of city temples. There was no indication of who had been found where, or by whom—corpses were just arriving stacked high on the backs of ancient flatbed trucks and being unloaded at the front gates to be sorted and perhaps eventually claimed. As they came off the vehicles their faces were photographed and the images stored on computers in the temple courtyard. Bearing in mind that this was a week after the disaster, the degree of bloating, discolouration and decomposition was extensive.

Families searching desperately for loved ones began their quest by pinning up their own photographs of the missing on a wall, accompanied by messages begging anyone who might have seen them to get in touch, in the hope that maybe they might still be alive in a hospital somewhere. Then relatives would go the temple courtyard and sit at the computers clicking through the hundreds of photographs of unrecognisable decomposing corpses, looking for the face of their son, daughter, mother, father, husband or wife. It was chaotic, hugely distressing and grossly ineffective. In the early days, bodies were being released to people purely on the basis that they thought they recognised them from one of these pictures. Not surprisingly, when a more scientific system eventually kicked in, many of the bodies were found to have been incorrectly identified and had to be recalled. This is, obviously, something to be avoided at all costs.

When we reached the temples we did three things immediately. First, we ordered refrigerated truck units to get the bodies cooled down and halt decomposition. Next we put a stop to families viewing galleries of decomposing faces. Then we suspended the release of any further bodies on hold until identity could be confirmed scientifically.

Before the 'reefers'—the refrigerated units—arrived, the bodies were laid out side by side in the temple forecourts. Attempts were made by the local teams to shade them from the direct heat of the sun by erecting tent-like coverings. At one stage, dry ice was packed around them to try to keep the temperature down. This never worked because the bodies closest to the edges suffered freeze burn and when our teams touched them, they got burned as well. The stench was indescribable. As the days wore on, the corpses continued to bloat and the turgidity caused by captured gas and fluids resulted in the pitiful sight of elevated limbs. When you looked down a long row of the dead, it seemed as if the raised arms or legs were trying to attract your attention. There was little running water, the heat was suffocating and the flies and rodent activity were of plague proportions. If Dante's hell were to be experienced on earth, those early days came very, very close to it.

Nobody was to blame for the situation. The first days after any disaster are often the most taxing and confused, and given the scale of this one, the practical difficulties were bound to be severe. It was the Norwegians who eventually came to the rescue by offering to fund a centralised temporary mortuary. Its construction, though, would take a while and in the meantime, it was a matter of coping as best we could with the limited resources available and a huge amount of lateral thinking and improvisation. And, tough though it was, this is the stage of an investigation I love the most: the hiatus before bureaucracy and politics step in, where experience and ingenuity come to the fore. This is the time when you feel you can make the greatest difference. I love the challenge of getting systems up and running. Once all is operating smoothly, I become easily bored. I believe we made a difference in those initial days in Thailand, although the red tape did start to appear very quickly.

Teams from the UK and other countries would stay on in the devastated region for close to a year, trying to identify and name those who had lost their lives. Many bodies were successfully returned to their relatives but a few stubborn identities remained unknown. The death toll was 5,400 in Thailand alone, and in some places whole families had been wiped out, leaving nobody to report them missing and nobody to provide the antemortem information that would allow us to make a comparison. In others, entire communities were razed to the ground, along with the records of their inhabitants, who were never missed or mourned. A memorial wall was built to commemorate all of Thailand's dead, and for some this remains the only epitaph. For DVI, it was a revolutionary operation that showed what can be achieved when international teams and governments work together.

Back in the UK, the meeting to which I had been invited to discuss the UK DVI response took place at Admiralty Arch. Slightly delayed by the London traffic, I found everyone there waiting for me when I arrived, which was not an auspicious start. In attendance were representatives of the highest levels of government, police and science policy. Some faces were familiar; others looked a little less welcoming. It was all rather stilted and awkward and I stood out like a rhino in a petting zoo. It was clear that the government had me pegged as a troublemaker and that officials had been briefed to be conciliatory. But my old friend from the Met was there, smiling encouragingly at his CW, so I knew I had at least one genuine ally in the room. And actually it turned out to be a really positive meeting. Everyone seemed to be in agreement that we needed a national response capability and that it should include police, government and scientific support. It was accepted that it was a matter of when, not if, we would next need to mobilise it. At last!

If any proof of that were needed, it would not be long in coming. Indeed, never had such a discussion taken place at a more timely moment. This meeting was held in February of 2005, a year that was to bring national and global deployments we simply could not have anticipated. The 7 July terrorist attack on London's transport network in the morning rush hour, which brought the city to its knees, was swiftly followed by more suicide bombings in the Egyptian resort of Sharmel-Sheikh. Hurricane Katrina, which devastated the US Gulf Coast in

August, and the Pakistan earthquake in October both struck while we were still trying to address the problems in Thailand. So it was a landmark year for DVI, not only within the UK but across the world.

Finally, in 2006, a national team of forensic experts, police, intelligence officers, family liaison officers and other DVI personnel was set up under the command of Detective Superintendent Graham Walker to undertake and co-ordinate the task of identifying British victims involved in disasters at home and abroad. It was clear that a UK DVI training programme was required to ensure that a police officer deployed from Devon in the south of England would work to exactly the same protocols and procedures as an officer from Caithness in the north of Scotland. This doesn't sound like it should have been an insurmountable problem but, as one officer commented to me, 'We have over forty police forces and they can't agree on a uniform, let alone a way of working.' Dundee University successfully tendered for the job of providing DVI mortuary training for police across the UK and, between 2007 and 2009, over 550 officers from every force in the country would come through our doors to learn the procedural and scientific underpinning of DVI.

◊

Disaster Victim Identification is neither rocket science nor brain surgery. It is, in principle, a very simple matching process. A family contacts the emergency telephone number given out by their government and states why they believe, or fear, that their loved one has been involved in the mass-fatality event. The individual will then be classified according to the likelihood that he or she actually has been involved. So in Thailand, for example, someone known to have been staying at a hotel destroyed by the tsunami would have been rated a more likely potential casualty than someone on a trip round the world who may or may not have been in the country and who simply hadn't been in touch with friends or relatives for a couple of days.

Prioritisation categories are important. It is just not possible for the police to give equal priority to every report they receive and there has to be some system to put those with the highest likelihood of involvement at the top of the list. These days, when everyone has a mobile phone, police and other authorities will receive multiple calls from the moment a major incident occurs. Following the 2005 London bombings, several thousand phone calls were made to the casualty bureau. And at one point in the aftermath of the Asian tsunami, 22,000 British citizens were reported to have been in the affected areas at the time. The eventual death toll for the UK was 149.

A family liaison officer with DVI training is sent to interview family and friends, first of the highest-priority potentially missing persons, and to record as much personal information as possible about them—height, weight, hair colour, eye colour, scars, tattoos, piercings, details of their GP and dentist and so on. They will find proxy sources of fingerprints within the individual's home, arrange for DNA samples to be taken from Mum, Dad, siblings and other genetic relatives and may even source DNA from items belonging to the missing person. This can only be distressing for a family in turmoil but the liaison officers will aim to collect everything they can, and probably more than they will ever need, in a single visit so that they do not have to put the family through further pain by returning to gather more, as this may sometimes erode confidence and weaken the relationship between the families and the authorities.

All this data is transferred on to a yellow AM (antemortem) DVI form. When the disaster has taken place abroad this will be sent, along with the DNA samples, fingerprints and dental charting, to the country in which the postmortem team is operating. In the mortuary there, they will be collecting the same information from victims and recording it on pink PM (postmortem) forms. I remember one bright spark asking me during training if we filled in the yellow forms in the morning, because they had AM written on them, and the pink ones in the afternoon. Sometimes it is not only the work itself that can be challenging!

In the matching centre, teams will bring together all the data from both forms. Matches are ideally made on the primary identifiers, but secondary methods may need to be used to corroborate an identification when DNA, fingerprints and dental information are either not present or inconclusive. The process is slow and quality control is critical. If we make a mistake, we will be depriving two families of their loved ones. Better to take our time and not make mistakes, even though we know criticism will not be long in coming when identifications are not confirmed swiftly.

The Dundee training programme was unique for its time. Having secured the contract in January 2007, we were aware that to run a course we were going to need to write a textbook to support the learning, and in very short order. We had a twenty-one-chapter text written and published by Easter (thank goodness for Anna Day and Dundee University Press), a copy of which was given to each officer. Our online distance-learning programme was also up and running by Easter. It would look a bit dated now, in the days of MOOCs (massive open online courses, designed for unlimited participation), but for its time, it was cutting-edge. Officers could gain access to the programme via their computers, anywhere and at any time convenient to them. Based on the textbook, it had twenty-one sections, each of which had to be completed in sequence before access was given to the next. After working through each section, officers had to sit an online multiple-choice test (or multiple-guess, as our students call it). If they answered more than 70 per cent of the questions correctly, the next section would be opened to them. If they didn't pass they could retake the test (a different one each time) until they did. When they reached the end, there was a further test covering the whole course. The critical information had by now become so embedded that virtually everybody passed this first time. Reinforced learning is a wonder to behold.

Only when the officers had completed the theoretical component of the programme were they allowed to come on a practical training week, where we would simulate a mass-fatality event. Our scenario was that a cruise ship carrying mostly retired people had run aground in bad weather on rocks off the east coast of the Outer Hebrides. Because of the infirmity of many of the passengers, a significant number had not survived. We had the permission of HM inspector of anatomy and of those who had bequeathed their bodies to us to use our anatomy cadavers as part of the training programme. This was the first time anywhere in the world that any DVI response force had been trained on real dead bodies and it was a sobering experience for all of our officers. They left with a new admiration for the Tayside donors and some even asked if they could come and pay their respects at our anatomy department memorial service, which they attended in full dress uniform.

The officers learned how to log a body from storage into a temporary mortuary; how to photograph, record and inspect personal effects and the body itself, and how to take fingerprints and retrieve other information that could be used to confirm an identity. They were taught the roles of the forensic pathologist, anthropologist, odontologist and radiographer. They filled out all the complicated sections of the pink postmortem INTERPOL DVI forms in the mortuary and trawled the multiple yellow AM forms—filled out by us—to try to find tentative matches. They were then required to present their cases to a genuine coroner or procurator fiscal, as if giving evidence at a real inquest, and justify their level of confidence in their identification.

The sense of camaraderie was superb and our interactions with so many officers from different forces led to many memorable moments—some poignant, some funny and all invaluable. The practical week included an assessment where each team was graded on their performance of various tasks. Some groups were subjected to mobile phones going off with horrendously loud and irritating bagpipe ringtones. The temptation for any red-blooded English person is to turn off the phone as soon as possible but the right response is to find it swiftly and to try to make a note of the number of the person calling. You will then be able to ring them to ask for the name of the person they were trying to contact, which might well help to establish the identity of the deceased more swiftly.

The officers were not allowed to handle the phone until it had been forensically tested so we would tease them by ringing the number and then hanging up as soon as they found it. Then we would wait until it had been placed in an evidence bag and ring it again. We had fun watching them scrabbling to write down the number before it disappeared. Frustrating it may have been, but it was effective in improving their responses.

We also placed rogue personal effects in pockets that were extremely unlikely to belong to the deceased person—items relevant to the wrong sex or wrong age, such as a lipstick in a man's trouser pocket, or a comb when the man was clearly completely bald. After the officers got wise to some of these curveballs, having heard about them from colleagues who had completed the course before them, and began to get a little cocky, we had to become even more creative, laying traps so far out of left field that they'd need binoculars to spot them. You cannot train people for the unexpected, but you can introduce the unexpected into their training.

With one group we planted a dummy hand grenade in a

body bag. This was not very plausible in the context of our retirement cruise scenario, but that was not important: the point of the exercise was to introduce distractors and assess response. We sat back and watched. When they found the grenade and alerted us, we set off every klaxon and bell we had, and amid the din everyone had to evacuate the mortuary. Initially they thought this a rather lame stunt and remained nonchalant as they congregated outside in their white scene-of-crime suits, waiting for 'bomb disposal' to arrive. They should have given us more credit for our deviousness. As time wore on and they waited and waited, they started to get antsy. They were on a time limit and they knew that they could not cut corners on the quality of the data they collected because they were being assessed. Before long they were hopping from foot to foot, tapping their watches and sweating.

When we decided we had kept the officers hanging around long enough, we gave the all-clear and allowed them back into the building. They set off at a great pace, moaning and groaning, in a rush to return to finish the job in hand within the allocated time. There was nothing lackadaisical about their approach now: they were focused and they were stressed. Time for one more little lesson. Mike, our senior mortuary manager, stopped them in their tracks and asked them where they thought they were going. 'Back into the building,' they chorused.

'I don't think so,' he said. 'You are all wearing contaminated suits. They will have to come off before you can go through a clean area.'

There were howls of protest ('But I only have my boxers on underneath!'). Oh dear. We cracked a wry smile while forty previously self-satisfied police officers stripped off outside the building and dashed through the front door in their underpants and T-shirts (fortunately, nobody was going commando, or we would have had to have rethought that tactic to spare their blushes, and ours). But they never underestimated us again. And it reinforced the message that the only predictable thing about a mass-fatality event is its unpredictability. These were the officers who would be deployed to air crashes, train crashes, terrorist incidents and natural disasters, and they would have to function efficiently and professionally while witnessing terrible events and traumatic aftermaths. They took it all with good grace in the end—after we bought them beer in the bar later.

For the third part of their course, which would earn them a university postgraduate qualification, we asked them to research any mass-fatality event in history of their choosing and then write an essay commenting on which aspects of the DVI component worked well, which did not and what they would or could have done at the time to improve on them. You should have heard the uproar. Why on earth did they need to write an essay? They weren't at school any more. But afterwards there was a general appreciation that the exercise was actually very worthwhile. It consolidated what they had learned through reading and practical exercise, and this critical evaluation brought them a valuable academic qualification in a subject to which they were already dedicated. They were tremendous sports, and many still talk about the course with fondness.

Some of the officers' essays were so well researched that we decided to use them in a second textbook—*Disaster Victim Identification: Experience and Practice*—donating all the royalties to the police charity COPS (Care of Police Survivors). Mark Lynch, of South Wales Police, and I wrote a chapter on the Aberfan disaster of 1966. This tragedy, along with Piper Alpha and the *Marchioness* collision on the River Thames, was a popular choice among the officers for their essays (we did have one on Vesuvius—not much scope for analysing DVI there!).

Aberfan emerged not only as a particularly good example of excellent working practices for its time, but also as a perfect illustration of how a job can be done well without the need for sophisticated modern technology. It set standards that would still pass muster today and it touched the hearts of every officer who chose to write about it, especially those from mining communities. It serves as a reminder that DVI is not a new process; that we are following in the footsteps of those who have gone before us, people who have dealt with the terrible tasks they faced with practicality, efficiency and compassion.

The disaster was caused by the collapse of one of the colliery spoil tips on a mountainside above the small mining community of Aberfan in south Wales. Tip number 7, which consisted largely of 'tailings'—the minute particles that remain after filtration—had unwittingly been positioned on top of an underground spring. On the morning of 21 October 1966, with the spring swollen by several days of heavy rain, more than 150,000 cubic metres of saturated debris broke away from the spoil heap and flowed down the mountainside at speeds reaching 50mph. At 9.15am, as pupils and teachers at Pantglas Primary School were settling down to their lessons on the last day before half-term, a massive wave of coal sludge crashed on to the building, burying it under nine metres of slurry.

Police and emergency services reached the school by 10am and every miner in the area, alerted by the sirens, grabbed his tools and headed off to help. When they arrived they found villagers, many of them parents, already digging into the slurry with their bare hands to try to reach the children. This was the first mass-fatality event ever to be filmed in real time: by 10.30am, the BBC were broadcasting live from the scene and the press were gathering. One rescuer remembered: 'I was helping to dig the children out when I heard a photographer tell a child to cry for her dead friends, so that he could get a good picture—that taught me silence.' Revisiting that testimony, I was reminded of my time in Kosovo.

Police from Merthyr Tydfil were there promptly and took charge of the search-and-rescue operation. This is the phase of a disaster where the living must take priority over the dead and it could take minutes, hours or days, depending on the nature of the event. In modern times, it is at the start of the second phase, body recovery, that forensic anthropology first becomes involved.

At Aberfan a medical reception was set up in the Bethania Chapel, 250 yards from the school. But with nobody found alive after 11am that day, the chapel quickly became a temporary mortuary. The vestry was used as a base for the army of volunteer workers and the Missing Persons' Bureau and to store 200 coffins. Postmortem autopsies were not required as the cause of death was known to all, but there was, of course, a desperate need for the bodies to be identified. The coroner and his officer worked with two local doctors to certify the deaths and liaised with surviving schoolteachers to piece together a list of the pupils most likely to be among them.

Each body pulled from the slurry was transferred by stretcher to the Bethania Chapel, booked into the temporary mortuary and assigned a unique reference number, which was pinned to the clothing as a label and remained attached to the body throughout. The URN was recorded on a card referencing system, together with details of whether the body was male or female, adult or child. The 116 dead children were laid out on the pews and covered with blankets—boys on one side, girls on the other. Three teachers helped with the preliminary identifications and a mortuary assistant washed the faces of the deceased to permit visual identification. Relatives queued patiently for hours outside the chapel, waiting for their turn to go in, one family at a time, to reclaim their loved ones. Once an identification had been confirmed, the body would be taken to the smaller Calvinistic Methodist Chapel to be stored until it was released for burial. The process proved difficult in only fifteen cases, due to extensive injuries. These victims were eventually identified by dental records.

Action in the immediate aftermath of a disaster is rightly focused on survivors, body recovery, identification and prevention of further fatalities, but in the long term, its legacy will endure for as long as there are victims to remember it, or a society that cares. Over fifty years after Aberfan, many of those affected still live with symptoms of post-traumatic stress. At the time, in the stoic culture typical of working-class communities, survivors were expected to 'just get on with it'. Nowadays, in an era that recognises the value of counselling and therapy, we understand that such suppression of trauma can have lasting effects on health and wellbeing. DVI practitioners are also more aware than ever of the need to try to avoid causing any more pain to survivors and bereaved relatives than is absolutely necessary.

This was an imperative that was acknowledged after the *Marchioness* disaster of 1989 by Lord Justice Kenneth Clarke, who chaired an inquiry into the identification procedures authorised by the coroner. His report, published in 2001, made thirty-six recommendations and suggestions for improvement and led to a review of the centuries-old coronial system. The biggest change was the introduction of a new police role, the SIM (senior identification manager), who would hold overall responsibility for the identification processes.

The victims of this tragedy had been at a birthday party aboard the *Marchioness*, a pleasure boat, on the River Thames when it was struck twice by a dredger, the *Bowbelle*. In the second collision the *Marchioness* was pushed underwater. Those trapped below deck had little chance of survival.

It took two days to recover the bodies from the water and the boat. They were transferred initially to a police station, where twenty-five of them were released to families after being identified visually by relatives or close friends. The coroner directed that families should not be allowed to view the corpses of the bodies that had been in the water the longest as putrefaction had progressed. Instead these individuals would be identified through fingerprint comparisons (DNA profiling was, of course, still in its infancy then), dental matching, clothing, jewellery and physical characteristics. All of them were to be given a full autopsy. Today we would question the necessity for this since, as was the case with the Aberfan victims, the cause of death was not in any doubt.

To aid fingerprint identification, the coroner for Westminster gave permission for the hands of the victims to be removed at the wrists where necessary. This decision was strongly criticised by Lord Justice Clarke and unfortunately became the focus of the entire DVI process. Permission had to come from the coroner as, under common law, it is only he or she who has a right to possession of the body. The hands were subsequently removed from twenty-five of the fifty-one deceased. Only when all of these bodies had been identified, which took nearly three weeks, were the remains released to their families. Many expressed their distress at not being allowed to view their loved ones. As well as adding to their anguish, it led them to question the certainty of identification in some instances and to a general mistrust of the authorities. The families pushed hard for a public inquiry, and eventually, in 2000—eleven years after the disaster—their campaign was successful.

Of the many important recommendations to come out of the inquiry, several arose from the unnecessary removal of the hands and the unwillingness of the authorities to let relatives decide for themselves whether they wished to see the remains of their family members. In three cases, the hands had not been returned with the rest of the remains. One pair had been found in a mortuary freezer in 1993, four years after the disaster, and had subsequently been disposed of without the knowledge or permission of the next of kin. Because the bereaved families had not been able to view their loved ones, many had been unaware that invasive postmortem examinations had taken place.

To receive this news a dozen years later came as a shocking blow.

A further recommendation by Lord Justice Clarke advocated strongly that families should receive only honest and accurate information and that it should be given regularly and as early as possible. Charles Haddon-Cave QC, who represented the Marchioness Action Group, said: 'Much goes on for understandable reasons behind closed doors. For this reason, there is a special responsibility placed on those entrusted with this work and the authorities who supervise it to ensure that bodies of the dead are treated with the utmost care and respect. That is what bereaved and loved ones are entitled to expect and what society at large demands.'

Our DVI training course gave us a great opportunity to discuss with police officers what we could learn from the past, what good (and questionable) practice looks like in difficult circumstances and what could be done better today. If Aberfan was an example of good practice, it was the *Marchioness* disaster, and in particular the decision to remove victims' hands, that had the most impact in terms of absorbing the lessons of previous mass-fatality events. Of all Lord Justice Clarke's statements, the words now branded most indelibly on every DVI operative are: 'It should be made clear that the methods used for establishing the identity of the deceased should, wherever possible, avoid unnecessary invasive procedure or disfigurement or mutilation and that body parts should not be removed for the purposes of identification except where it is necessary to do so as a last resort.'

Forensic practitioners have always tried to do what we thought was for the best, sometimes in an effort to protect families from harrowing sights. If a body has started to decay, or has been damaged and fragmented by fire or explosion, in the past we would perhaps have advised them against viewing it. But we have no right to make that decision for them, let alone to impose restrictions on what they can and cannot see. We do not own the body. It is impossible in any case to predict how families and friends will respond to the empty shell of someone they love, regardless of its condition. So if a mother wants to see her dead child and hold his hand, if a husband wants to kiss the remains of his wife or a brother wants to spend a last moment in peaceful silence with his brother, all we can do is prepare them for what they will face and be on hand to help.

Whereas Aberfan was what we would today call a 'closed' incident, in which the names of the deceased were known and nobody was left unaccounted for, the *Marchioness* disaster was classified as an 'open' incident, where DVI becomes more complex, and therefore more difficult to manage. These are situations where it is not known at the outset who the deceased might be or how many people might have died or have been injured, as often many of the surviving casualties may be in too serious a condition to be able to confirm their identities. The *Marchioness* had no definitive passenger list and the number of people missing was initially unclear.

When an 'open' incident is the result of a terrorist attack, priorities may alter. While DVI procedure remains exactly the same, the process of body and evidence recovery may be very different. In the attack on London's transport network in July 2005, suicide bombers detonated three bombs in quick succession on separate Underground trains in different parts of the city and a fourth aboard a double-decker bus. Fifty-six people lost their lives, including the four bombers, and 784 were injured. While identifying the dead victims was of course urgent, the first priority was to help the survivors and the second to identify the perpetrators.

This may sound strange, but in dealing with the UK's first-ever so-called Islamist suicide bombing, the authorities were following an accepted protocol for mass-fatality events caused by acts of terrorism. It is vital to establish whether the perpetrators have been killed in the incident and to be able to trace their networks in case it is part of a co-ordinated event, so that everything possible can be done to prevent further fatalities. The 2005 London bombings did indeed prove to be a co-ordinated attack, although unfortunately in this case the explosions occurred too close together for any of them to be stopped.

In 2005, the UK was nowhere in the hierarchy of international DVI, but by 2009 we were leading the world. I could never have imagined that my somewhat intemperate missive to the government after the Asian tsunami would have the effect that it did, or that it would go on to play its part in establishing our DVI response force. And I am very proud of the fantastic work that UK DVI went on to perform and still carries out today to the highest international standard.

I cannot state it often enough because it really is something that I believe in my very core: we must never forget that with disasters it is not a question of 'if', but 'when'. So it is vital that when the next one strikes we are ready to respond to the very best of our abilities, however big or small the incident may be. In our world of increasingly senseless acts of violence, the UK's first DVI commander, Graham Walker, reminded us all that whereas terrorists only need to get lucky once to accomplish their mission, our investigative forces can't rely on luck—they have to win every single time to keep us safe. With the best will in the world, that is unrealistic and so we must train and prepare for all eventualities, while constantly praying that we will never need to put them into action. But when we do, our response should demonstrate that our humanity transcends the worst malevolence of which our species and nature are capable.

CHAPTER 12

Fate, fear and phobias

'Men fear death as children fear to go in the dark'

Francis Bacon, philosopher and scientist (1561–1626)

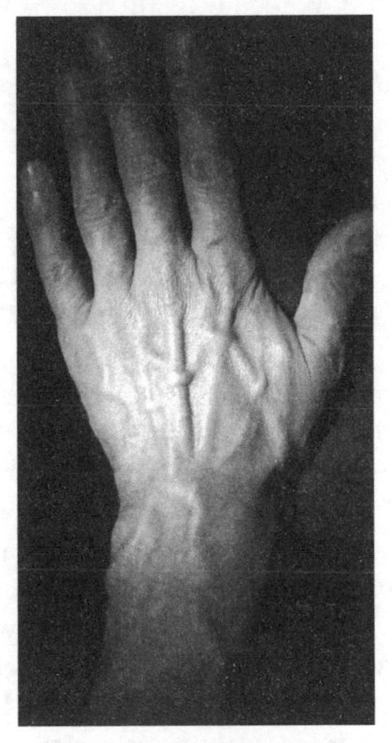

The anatomy of the hand: variation in veins.

I share my passion for identification of the human with my dear friend and mentor Louise Scheuer, who was responsible for me getting my first proper tax-paying job at St Thomas' Hospital in London. While I was still writing up my PhD, Louise phoned me one day to say there was a vacancy for a lecturer in anatomy at St Thomas' and that she thought I should apply.

Nobody was more surprised than I was when I got the job. The chair of the panel, an august professor of neuroscience, clearly wanted someone with a biochemistry degree and was very dismissive of this 'unqualified anthropologist'. I know that it was the last question, posed by Professor Michael Day, the head of the department, that sealed the deal. He asked me whether I would be capable of going into his dissecting room that afternoon and teaching the brachial plexus. I said that of course I would—and the job was mine. I have since used this ploy myself many times in interview panels but, having learned from my own experience, I have taken it that little bit further. I ask people applying for a job in anatomy at Dundee the same question, and when they say that of course they could teach the brachial plexus, I invite them to draw it. The brachial plexus, in case you are wondering, is a bunch of nerves that run between your neck and your armpit and looks a bit like a plate of spaghetti. Mean, isn't it? I am glad Michael didn't do that to me or he would have found me out. I could draw it now, but I couldn't have done it then.

The man who considered me unqualified eventually became my boss when our departments at Guy's and St Thomas' merged a few years later. I had the pleasure of teaching anatomy in his department for a number of years, but I don't think he ever forgave me, because when he thanked me at my leaving do in 1992 for all I had done, he called me Sarah. I clearly made no lasting impression there. But at St Thomas' I was lucky to have a group of truly wonderful colleagues who are still friends to this day. Most importantly, it was the beginning of a long and productive partnership with a woman who has forgotten more anatomy than I will ever know and who has been my friend, my inspiration and my teacher for over thirty years.

When Louise and I set up the UK's first forensic anthropology teaching programme in 1986, she had a persistent and irritating moan. Every time we had to analyse the skeletal remains of a child she would say, 'Why isn't there a textbook to help us?' I would respond by suggesting she wrote one herself and she would tell me to behave myself. She has the most wonderful ability to sound like a governess addressing an errant child ('Oh, come on, for goodness' sake!'). After about four years of this exchange, I decided to rebel and changed the record. 'Why don't we write one together?' I proposed.

And so began our biggest-ever writing project. We wanted to produce a textbook on the development of the human skeleton which examined every single bone in the human body, from the point at which it first forms until it reaches its full adult state. It was never going to make us rich or appear in the *Sunday Times* bestsellers list, but we felt it was essential to fill a glaring hole in forensic anthropology's academic arsenal. Since there was no other book in existence that considered the detail of the child's skeleton at the level we needed, we were starting with pretty much a clean slate.

It would take us nearly ten years. First we had to accumulate all that had been written and published elsewhere on the subject in the last 300 years or so. Then we had to identify specimens that would exemplify what we wanted to illustrate and, where there were gaps in the knowledge, conduct our own research. It became crystal clear to us very quickly why this had not been attempted before: it was going to be a labour of love, and a very slow one. Indeed, it would come to dominate our lives for the duration.

We finally saw our work published in 2000. *Developmental Juvenile Osteology* is a very, very long book and not exactly a page-turner, given that there are over 200 bones we needed to consider, but it was a really fascinating and rewarding one to write and it became a hugely significant marker in our professional lives. I loved those moments when I would get an excited call from Louise saying, 'Did you know . . .?' or 'I finally understand why . . .'. There were many of these delightful discoveries, some of which blew our own theories out of the water as we learned together and, very slowly, started to tie it all into a magnum opus of which we were both hugely proud.

By 1999, when I was doing my first tours in Kosovo, we were close to finishing although there was one illustration that vexed us terribly. We were unable to find any specimen that showed the growth centre at the inferior angle of the scapula (shoulder blade).

I must admit that on one occasion it was Louise, rather than Tom or the girls, who got my rare and precious satellite phone call from Kosovo. In the makeshift mortuary at Velika Krusa, I had just seen a specimen that exemplified exactly what we had been searching for. We were both ridiculously thrilled. I was given permission to photograph it for use in the book but unfortunately, I had failed to realise that while all the other pictures we had were of nice, clean, dry bones, in this one there was still some tissue. I am glad now that the illustrations were not in colour as it would have stood out as a bit gruesome. But the image was priceless from the educational point of view.

By the time the book was completed, I had been back in Scotland for several years, Louise had retired and my second season in Kosovo was looming. At that point, Louise and I were probably better versed in age-related changes in the child's skeleton than anyone else on the planet. My grandmother, who passed down to me her belief in fate, used to say that sometimes there is a reason why we find ourselves in a certain place at a certain time and often it has nothing to do with our own plans, choices or desires. We are there because fate has put us there, quite possibly to help someone else. That I found myself in Kosovo at that precise moment in my life when I had all this knowledge at my fingertips was, I am convinced, predestined.

One of the indictment sites assigned to us in 2000 involved the murder of virtually an entire family. During the war with the Serbs, those Kosovo Albanians who could stay out of the towns and villages tried to do so, to keep clear of the Serb forces, which tended to be most active in the more populated areas. On a March morning in 1999, this family had been making a trip from out in the countryside to the nearest village to pick up supplies, with Father driving the tractor and everyone else perched on their flimsy wooden trailer behind him. Without warning, the trailer was struck by a rocket-propelled grenade (RPG) launched from the hillside and blown to pieces. On board were eleven of the man's family: his wife, her sister, their elderly mother and his eight children. The youngest was just a babe in arms and the eldest were twin fourteen-year-old boys. None of them survived.

As the man clambered down from his tractor, now separated from the trailer by the blast, a sniper shot him in the leg. Injured and bleeding, he was able to crawl into the undergrowth for cover. He tied his belt around the wound to stem the flow of blood and, in the certain knowledge that all of his family were dead, waited for the light to fade and silence to return, desperately hoping that, even if the snipers had not gone away, they would not be able to see him so clearly in the twilight. He knew that if he did not recover what was left of his loved ones, they would be preyed upon by roaming packs of wild dogs and he could not allow that to happen.

When he felt it was safe to crawl out from under the bushes, he began to search for the remains of his family. The RPG had fragmented them all, apart from the baby, who was still perfectly intact. It is testament to the savagery of the hit and the enormity of his grim undertaking that he was not able to retrieve all the parts of their bodies: he told us that he had only been able to find the right side of his wife and the bottom half of his twelve-year-old daughter. As my husband wondered, how can anyone summon the presence of mind and the fortitude to do what he did? Where do you find that depth of courage, strength and commitment to those dearest to you? Tom, quite understandably, reflected that it would have been too much for him to want to go on, and he would probably have ended his own life there and then. But this man didn't. His determination to search the grassland in the failing light for the bloodied remnants of his family, growing weaker and weaker all the while through his own blood loss, is truly remarkable.

When he had gathered together all he could, he buried their remains, using a spade salvaged from the wreckage. He chose a distinctive tree as a marker for their resting place so that he would know where they were located and could find them again when he was able to return. Having toiled for hours, this tormented man's last act was to place the body of his baby son on top of the accumulated fragments of the rest of his family, cover them with soil and pray for their souls.

It was over a year later that the ICTY investigators identified this as an indictment site for the case being built against Slobodan Milosevic and his senior officers. They believed the attack was a clear act of genocide as the deliberate targeting of a man and his whole family could not be justified as a legitimate act of war. The man, who had somehow survived, took the investigators to the tree where he had buried his loved ones and gave them permission to exhume the bodies. Not only did he want justice for his family, and the families of other Kosovar Albanians, he also feared that, because their body parts were all mixed up, his God would be unable to distinguish between them to find their souls. He could not be at peace until he knew they were safe with God and he was desperate for each to have their own named grave, so that their souls would be recognised and rescued from the cruelties of the world.

I was not present at the exhumation but I was aware of the monumental task that awaited our team. We had to try to identify and separate the commingled remains of eleven severely fragmented and decomposed individuals, eight of whom were children, to a standard that would meet international evidential admissibility, while at the same time being mindful of the needs and wishes of a brave man who had lost everything.

At the mortuary, we expected a delivery of eleven body bags but there were only enough body parts to fill one and a half. This was the sum total of what this man had been able to find and bury that terrible day. The remains were very badly decomposed and although some soft tissue survived, most was little more than a liquefied mass interspersed by bones. Examining them was an extremely difficult and painstaking task, not to say an unpleasant one. There was no point in the whole team sticking around just to stand and watch, so we decided to give them a welcome day off and I stayed on with the mortuary technician, the photographer and the radiographer to see what we could achieve.

We laid out twelve white sheets on the floor—one for each of the potential deceased, labelled only with their reported ages, and one for the residue that we would inevitably not be able to assign with any confidence to a specific individual. Even DNA in this situation would not have helped us as the dead were all members of the same family and we had no reference DNA for comparison purposes. Even if we'd had such samples, the probable cross-contamination of body parts that had been buried together would make the chances of extracting reliable source DNA very slim. So this had to be an old-fashioned anatomical identification process—and, with so many children involved, at that moment in time, Louise and I were probably the most experienced forensic anthropologists available anywhere to undertake the task. At that point I was the one in Kosovo and Louise was back in London, but she would be at the end of a phone if I needed her, and that shored up my confidence no end.

We started by X-raying the two body bags to make sure they contained no unexpected ordnance. The images were a sobering sight indeed—shadows of jumbled pieces that would become a grim and taxing human jigsaw. The first body bag was opened and, lying on the top, still in his blue sleep suit, was the baby. Although he showed some quite extensive decomposition he was still reasonably intact and could be laid directly on to his own mortuary sheet in the certainty that he was the child aged six months.

With the rest we had to remove the remains bone by bone, cleaning away the adhering residue of decomposing tissue, identify the bone, assess its age and then place it on the mortuary sheet allocated to the corresponding individual. We were able to separate the women. The children's grandmother was identifiable through her lack of teeth and her advanced osteoarthritis and osteoporosis. The two younger women were more challenging but for one of them, most likely the elder of the two, we had only the right side, which corroborated the story of the survivor. So she was in all probability his wife.

When it came to the children, if we got our analysis right there would be no duplication in terms of their ages until we came to the fourteen-year-old twin boys, since none of the others were the same age as one another. The bottom half of the twelve-year-old girl was recovered and she was identified relatively easily. The remnants of the younger children, aged three, five, six and eight, were sparse, but there was enough there for me to be confident that distinct parts could be separated out from the general tissue mass.

We had something now on every sheet except the bones that would allow us to differentiate between the twins. All that remained of either of them were two partial upper torsos and upper limbs as far down as the elbows. We knew these body parts belonged to the twins as they were the only children of the right age—but how to tell them apart? One set of limbs was associated with a Mickey Mouse vest, so we asked a police officer and interpreter to determine from the father whether any of his children could have been wearing such an item. We did not specify that the child was one of the twins, or even that he was a boy. The answer came back naming one of the twins as a Mickey Mouse fanatic, which enabled us to tentatively distinguish between them.

It was a long day, nearly twelve hours with barely a pause, but by the end of it we had identified and assigned as much of the material as we possibly could and all eleven mortuary sheets allocated to the remains of specific victims contained some definite representation of them. We obtained a list of their names, based on their ages, from the father and started to pack the pitiful remains into separate body bags. When the authorities refused to allow us to release the bodies of the twins as named individuals, my SIO, Steve Watts, went into battle. We explained why there was no way to separate them with any more certainty than we had been able to achieve, we talked them through our rationale and we wore them down into submission.

It meant that as we handed over each body bag to the survivor, the interpreter was able to tell him the name of the person it contained. The interpreters played such a crucial part in enabling us to be accepted by the community and they did a truly remarkable job. It was they who had to talk to the families, take their statements and translate back to them what we had found, all the time trying to insulate themselves from the horrors they had to hear about and communicate every single day of their working time in Kosovo.

While I make it a rule never to get personally involved—I couldn't do my job properly otherwise—in this case I think we may have crossed the line by a hair's breadth in forcing the issue about the twins. But we felt a particular connection and responsibility for these identifications, perhaps partly because most of the victims were children and partly because their father had suffered so much and borne it all with such courage and stoic dignity. It was the least comfort we could provide, and we knew that it was not possible for any more sophisticated scientific testing to improve on our professional opinion.

As it was, we still had to try to explain through the interpreter, gently but honestly, why the bags were not full, and why there was a twelfth bag which contained a mixture of remnants. With incredible grace, he accepted it with a calm understanding that was almost otherworldly. It was an extremely affecting day and we were physically and emotionally exhausted. When he shook hands with us all and said thank you, we found it hard to comprehend just how he could thank us for the job we had just done. But as my granny used to say, fate isn't there for our convenience.

It weighed heavily on us that we had not been able to hand back a tidy set of eleven separate remains, which would have reassured this father that we had been completely successful in our reassignment. But this was not a humanitarian mission, this was a forensic investigation of war crimes, and if we were ever tempted to assign body parts just for the sake of neatness or consolation, we would be guilty of professional misconduct. We had to be certain that, if this human evidence were ever to be examined again in the future, the person we said was in each bag was genuinely the person we believed him or her to be.

We would never have achieved as much as we did without the knowledge and understanding of the juvenile skeleton I had gained from working on our textbook. That day I put into practice everything Louise and I had learned over those ten years of writing, and it brought home to me precisely why it had been such an important project for us both to undertake. I may have been the one in Kosovo that day, but Louise was inside my head all the while, reminding me of details as I checked and rechecked, kept notes, compiled lists and satisfied myself that I was as sure as I possibly could be before committing anything to a sign-off.

The task we undertook in Kosovo was a huge responsibility but it was enormously rewarding. What were the chances that I would just happen to be on hand for that particular case? Perhaps the attitude my grandmother would have taken is the right one: that it was never about chance, that all of it—my move to St Thomas', partnering Louise, the work on the book—had been leading up to that moment. And beyond, since our book might in the future also enable somebody else to bring some small comfort in another dreadful situation. I certainly feel that even if we never again need to call on all that information, it was worth it just for that important identification in a mortuary in Pristina in 2000, the year it was published. Every time I leaf through it and see that picture of the scapula, I think of the father of those children and our textbook remains in my mind a fitting tribute to that single case.

◊

One of the questions most frequently asked of forensic anthropologists is how we cope with what we have to see and do. In response, I usually joke that it involves large amounts of alcohol and illegal substances, but the truth is I don't think I've ever taken an illegal substance in my life and, other than the odd Jack or two, even drinking doesn't do it for me these days. Do I wake up in the night sweating? Do I find it difficult to sleep? Do scenes from my work replay over and over in my mind? The answer to all of the above is a rather boring and mundane no. If pressed, I have some stock explanations about the need to remain professional and unbiased, the need to focus on the evidence, not what it represents from a personal or emotional perspective, and so on, but to be perfectly honest, I have never been spooked by the dead. It is the living who terrify me. The dead are much more predictable and co-operative.

Recently a colleague in a very different field said to me, with incredulity in his voice, 'You talk of these things as if they are as normal as making a cup of tea. For the rest of us they are extraordinary.' Isn't it just part of life, though, that one man's meat is another man's poison? Perhaps forensic anthropologists are the sin-eaters of our day, addressing the unpleasant and unimaginable so that others don't have to. It doesn't mean, of course, that we don't have our own weaknesses.

None of us is without fear—it is, after all, one of our oldest and strongest emotions—and we are all afraid of something. During my career there have been times when I've had to confront my single, genuine, full-on phobia. It is one I have lived with since childhood and although I have done my best to cope with it, I have never conquered it. But true self-knowledge often lies in accepting our anxieties and flaws and facing up to our fears. Mine is my utterly ludicrous morbid fear of rodents. Any kind of rodent: mouse, rat, hamster, gerbil, capybara—all of them.

Very recently a local charity which supports our anatomy department was kind enough to give us a Christmas present: they sponsored a rat for us. Chewa (yes, he has a name) is a 3lb African giant-pouched rat who sniffs out tuberculosis for a living. He is a HeroRat and has saved over forty lives. Yes, I am impressed, I really am, but however much he deserves my admiration and affection, I am sorry—he is a *rat*!

It may seem strange for a forensic anthropologist, who deals every day with the dead, body parts and decomposing matter that would turn many stomachs, to suffer from such an irrational phobia. I agree, but understanding that is no consolation and doesn't make the rodent community any less terrifying. It has been a recurring issue for me throughout my entire life and, in many ways, I suppose it has even helped to shape my career choices.

It all began on the idyllic shores of Loch Carron, on the west coast of Scotland, where my parents ran the Stromeferry Hotel until I was eleven years old, when we moved back to Inverness. One summer the dustmen ('scaffies', as they are known in Scotland) went on strike and the black bags of rubbish started to pile up around the back of the hotel. It doesn't take long for the air in a thirty-bedroom establishment at the height of the summer season to turn sour and for our furry rodent friends to identify a source of free rancid food. I was nine, and I remember very clearly walking around the back of the hotel with my father one sunny afternoon when he quite calmly asked me to hand him a broom that was propped up against the wall. I complied without a second thought.

My father always vowed that what followed never occurred but, believe me, it did—I know it did, because it has haunted me every day of my life since and resurfaces whenever I have to interact with any furry rodent critters. He had seen a rat, which he proceeded to corner against a wall. I was horrified: to me it seemed enormous, and it was scared and squaring up for a fight. If I close my eyes, I can still see its shining red eyes, bared yellow teeth and lashing tail. I swear I heard it actually growling. I watched, transfixed by terror, as it leaped around, trying to escape, while my father beat it to death, terrified that it was going to jump on me and bite me. He thrashed it until the concrete turned red with its blood and it eventually stopped twitching. I have no memory of my father picking it up and throwing it into the rubbish. Perhaps I was too traumatised. But I never again walked round the back of our hotel on my own and from that moment onwards, I developed an unhealthy, deep-seated fear of and aversion to any and all rodents.

This phobia continued to be a problem when we returned to live in the countryside near Inverness. Our ancient, thick-walled house was sandwiched between a burn on one side and a field on the other. This meant that in the winter, all the nasty little scrabbly blighters used to come in from the cold to live off our heat and raid our larder. I would jump on to my bed at night for fear that a rat would jump out from under it and grab me by the ankle. As I lay in my bed I could hear them scampering across the rafters above my head. Suddenly, one would misjudge its leap and I would hear it fall and scutter down the space in the wall. Convinced that it was going to emerge in my room, I would pull the blankets up around my ears and tuck them around me so that there was no gap for these brutes to get in.

I always walked between my bedroom and the bathroom at night in the dark and in my bare feet. Imagine my fright when one night, as I padded along the landing, I stood on something furry and moving that squealed at me. I freaked out, and for months never left my bedroom at night, no matter how urgent the call of nature might become.

As a student, I was confronted with rats in my zoology class: a bucket of dead ones this time, which we were expected to dissect. I would have dissected absolutely anything else but nothing was ever going to persuade me to voluntarily touch a dead rat, let alone pick one out of a bucket. I got my dissection partner, Graham, to choose one for me and pin it out on the wax board. I made him cover its head and nasty little pointed teeth with a paper towel, and then place a second paper towel over its tail, because I couldn't bear to look at that, either. Only then was I able to cut open its thorax and abdomen with my scalpel, guddle around in its innards and take out a liver, a stomach or a kidney.

When it came to disposing of the corpse, Graham had to unpin it for me and drop it into the bucket (he was a good friend). Needless to say, I was never going to be a zoologist or, for that matter, any kind of lab-based career researcher. As I recounted earlier, when it came to my honours research project, the reason I went into the field of human identification was to avoid the prospect of handling dead rodents.

It was inevitable that this would become a problem at St Thomas' Hospital, given its location along the south bank of the River Thames. When I walked into my office on my first day there and saw the mousetraps and little bowls of poison along the walls of all the offices, I knew things weren't going to end well for me here. Inevitably there was going to be a close encounter of the furry kind at some point. It happened one morning when I arrived at my office, walked over to my desk against the window and turned to see a monstrous dead rodent lying in the middle of the floor. In truth it was only maybe four inches long, but as far as I was concerned it might as well have been the size of Chewa.

I rang our technician, John, and screamed at him down the phone to get up to my office immediately and help me. He tore upstairs, bless him, obviously thinking I had been attacked, and found me sitting on my desk, shaking, the tears rolling down my face. I pointed to the dead mouse and explained that there was no way I could step over it to get out of the room. It had me trapped me like a prisoner. He could have laughed and ridiculed me but he was such a lovely man that he just quietly took the mouse away and never mentioned it again to me or, as far as I know, to any living soul. In fact, I think he checked my office for me regularly because I never again found a mouse in it. I felt such a twit but the phobia was fully hard-wired by then.

And then there was Kosovo. Our mortuary at Xerxe, being a former grain store, was a magnet for rodents—hordes of them. Every morning I would plead nicely with our Dutch military security to open up for me and then to go into the building making a lot of noise to chase all the rodents away. I couldn't have crossed the threshold knowing they were there. I could hear the creatures scurrying along the pipes and squealing in complaint at being disturbed. The soldiers were very kind to me and never grumbled about doing this. Perhaps, seeing what I was prepared to deal with in the course of my job, they understood that I was not in general a complete nut job or wuss, and that my terror was real, if utterly illogical.

My worst experience, though, was at Podujevo, to the north-east of Pristina, where it was alleged that early in 1999 the Scorpions, a Serbian paramilitary organisation, killed fourteen Kosovo Albanians, mostly women and children. The bodies were said to be buried under the local meat market. It was a known ploy to bury bodies with a dead cow or horse on top of them, so that when we dug down and discovered non-human remains, we would assume this was simply the grave of an animal and would not trouble to dig any further.

The day we began to excavate under the meat market was swelteringly hot. We had a mechanical digger to help us and it slowly cleared away the topsoil, inch by inch, until something was spotted by the person assigned to the watching brief. I was standing away from the edge of the freshly dug hole, waiting in the shade of the opened boot of our car, when I heard a bit of a commotion. As I walked towards the hole to see what was going on, one of the soldiers called out my name with an urgency that stopped me in my tracks. Having attracted my attention, he held my gaze, pointed a finger at me and shouted: 'Stay! Don't look!' I did as I was told.

It turned out that the digger had hit the anticipated horse's carcass and in the process had disturbed a nest of rats that were using the remains as a food source. As the digger hit the nest, its inhabitants were leaping about in a frenzied attempt to find their 'rat runs' and get clear of the perceived danger. Only when they were all gone did the soldier give me the all-clear, smile and say, 'Go on, get in the hole, girl.' Yep, it stank. Yep, I was up to my elbows in decomposing horse gore. But there were no rats, so I was happy as a clam.

The soldiers looked after me and my sensitivities—I have no qualms about being a girlie when I need to be—but they didn't mollycoddle me and I respected them for that. Indeed, molly-coddling was in distinctly short supply in our team: eau de liquid horse is never going to be a cosmetic-counter best-seller but the stench in that hole was worse than anything I had ever encountered, I can tell you. When it came to lunch-time, I was very politely, but forcefully, asked to sit downwind on my own. Oh, the indignity of it all.

◊

Given the extreme nature of our work and the living conditions in Kosovo, it was inevitable that everyone's fears or vulnerabil-ities would be exposed at some point or other. We were all permitted moments when it was OK not to be able to cope. What was important was that, when it mattered, we looked after each other.

Any experience that brings you into contact with mass fa-talities or the inhumanity of the human race will, of course, leave its indelible imprint on your life. I have participated in a number of public events with the bestselling crime writer Val McDermid, and we have become good friends. Val is a very smart and perceptive lady and she has told me that on these occasions I can be outrageous and have an audience in giggles, or indeed bawdy laughter, yet the minute we start to talk about Kosovo, she senses that a veil seems to come down as I retreat to a distant place. She says that my tone becomes reflective and the atmosphere is tinged by sadness. I am never aware of that but it doesn't surprise me.

It is a subconscious response, originating, I suspect, from the need to maintain perspective. Compartmentalising, on the other hand, is a cognitive choice, something you train yourself to do. I don't believe I am uncaring and cold but I do consider myself level-headed. Hard as nails when I want to be, especially when I go to work, where I make every effort to keep visceral reactions and emotional investment to a minimum by opening an imaginary door into a detached, clinical box inside my head. If forensic experts allowed themselves to dwell on the immensity of human pain or on the gruesome spectacles we encounter, we would be ineffective scientists. We cannot take on the suffering of the dead. That is not our job, and if we don't do our job, then we help nobody.

The actor and advocate of communicating science Alan Alda says that sometimes the greatest things happen at thresholds, and it is by consciously stepping across a threshold envisioned in my mind that I move from one world into another. There are probably several very self-contained compartments lurking in there—I think of them as rooms—and I know them all so well that I automatically choose the one that best suits the job at hand that day.

If I am working with decomposing human remains, I find a room where smell doesn't register. If I am dealing with murders, dismemberment or traumatic events, then I spend the day in a soft space where there is a sense of calm and safety. If I have material to examine relating to child abuse I will take myself to a far corner of the room where there is little sensory connectivity so that I do not transfer what I am seeing and hearing in that alien landscape of incomprehensible violation into my personal space. While occupying each box I am aware that I am striving to be an inert observer, albeit one proactively applying scientific training to observation, and not necessarily a psychologically sentient participant. It is almost a form of analytical automation. The real me remains outside that box somewhere, removed and protected from the sensory bombardment of the work that goes on inside.

When I have examined, recorded, formed an opinion and completed my task, all I need to do is open the door of the room, step back across the threshold, lock it behind me and I am back in my ordinary life. I can then go home and be me—a mother, a grandmother, a wife, a normal person. I can sit and watch a film, go shopping, pull weeds in the garden or bake a cake. It is imperative that the door is kept locked and that I don't let anyone else inside the box to poke around, or in any other way allow one life to bleed into the other. They need to stay completely separate and each must be protected.

Only I know the access code to the door; only I know all of the experiences that reside within the box, and all of the potential demons that might be lurking there, trying to look over my shoulder while I work. I can live alongside those experiences comfortably when I inhabit their forensic world, but when I leave it, they must remain locked inside. I do not intend ever to release them. I feel no need to 'address' them or talk about them in counselling. I will never commit most of them to paper or record them in any way, other than in my forensic notes. In some instances, I am bound by confidentiality, but even when I am not, I hold myself responsible for safeguarding the vulnerability of others, living or dead, and not betraying their secrets. There are things, too, I have seen and done of which my family and friends simply do not need to know, and should not know. All of the cases I have discussed in this book are already in the public domain. Those that are not are incarcerated in their own private box.

It is also a matter of protecting myself. It is only realistic, in my job, to fear the prospect of a Pandora's box-type meltdown one day. If that door is not closed properly, and someone goes snooping around where they haven't been invited, there is a risk that some or all of the demons might escape. Fortunately, I have thus far managed reasonably successfully to keep my two worlds apart. If they ever collide, and I fall victim to the clutches of post-traumatic stress, I will stop doing the work I do, because I know I would no longer be effective as an impartial observer.

We must respect the potential effects of our work on ourselves and never underestimate the crippling nature of that clinical condition, which can manifest itself without warning and to which we must never consider ourselves immune. It could be triggered in any of us by some incident, big or small, that nobody could ever see coming. I have seen the devastating impact of post-traumatic stress on colleagues who have become so haunted by what they have experienced that they have been unable to work and how it has destroyed lives, relationships and careers. Good mental health requires regular attention, so we must remain vigilant and keep guard on those incarcerated demons. But whenever one or two of them do manage to escape, the havoc they cause cannot be attributed to any weakness on the part of their owner.

Since I believe there is nothing to be feared from death herself, in my case any such lurking demons are more likely to be associated with the crimes of those still living and breathing. I have only once been aware of my work threatening to creep into my private life and the trigger was probably the subconscious effect on my psyche of the terrible things humans are capable of doing to other humans, rather than any ghosts of the dead.

It happened when our youngest daughter Anna was invited to a school dance by a boy. She looked gorgeous in her long gown, with her hair styled into a grown-up 'do'. Tom and I were present at the big event as part of the posse of parental chaperones, watching to ensure that propriety was observed, that no alcohol was surreptitiously consumed and that the only thing smoking was the barbecue. At one point, scanning the dance floor for Anna, I saw her dancing with a middle-aged man I didn't recognise. It was a small school and I thought I knew everyone. Nobody was able to tell me who he was.

I felt my pulse start to race, my stress levels rise and a flush envelop my face. It took every scrap of willpower I could muster not to run across the dance floor and demand to know who he was and why he was dancing with my daughter. Forcing myself to remain on the sidelines, I monitored every single step of that dance. I watched where he put his hands as the two of them twirled and waltzed, gauged just how close they were dancing and scrutinised his physical interaction with Anna as they talked and laughed. The poor man never put a foot (or hand) wrong, but that did not stop the alarm bells going off in my head.

Acknowledging that my reaction was completely over the top, not to say way out of character, I slowly talked myself down and rationalised the situation back to normality, even though my heart rate wasn't listening. I told myself that we were at a well-organised school event, with parents and teachers everywhere, I was standing just a few feet away from my daughter and there was no sign whatsoever that she was in any danger. It didn't prevent me from going across after the dance was finished and asking her, as casually as I could, if she was having a nice time and, by the way, who was the man she'd been dancing with? It turned out that he was the father of the boy who had invited her. I felt such a numpty but I was definitely still rattled.

Is that what post-traumatic stress feels like? I don't know, but it was a sense of threat and panic I had never experienced before or, thankfully, since. You could put it down to the standard fears of an overprotective mother, but I knew that I was definitely out of sorts and this was not normal for me. It was a crazy moment but the fact that I caught it, and saw it immediately for what it was, reassured me that, should I ever be struck by post-traumatic stress, there is a good chance I would recognise it.

We had handled four paedophile identification cases that week, so maybe that could explain my uncharacteristic response. Although most of our work is, naturally, with the dead, the reach of forensic anthropology now extends to the identification of the living. One important innovative strand is the assistance my team at Dundee is able to offer, nationally and internationally, in cases of child sexual abuse. This came about as a result of our efforts to answer a question posed by a particular investigation.

Research into identification is often driven by such questions, which present us with an exciting opportunity, because, very occasionally, an entirely new field may be unveiled. With much of the early work on identification having been in use for over a century, it is a rare and wonderful event when something fresh comes along. The prime example of this is DNA profiling, where the system invented and developed by Sir Alec Jeffreys at Leicester University eventually became the forensic standard worldwide and changed our domain for ever. So much so that we often forget that we did not have DNA analysis in our forensic toolbox until the 1980s.

For me, such an avenue opened up when the police contacted us for help with a tricky case. Although neither the methodology nor the underlying principles to which we turned were new, the ways we found to apply them were. Sometimes it takes a change in societal circumstances to bring about the rekindling of a lost art or for a particular approach to become 'of its time', and that was indeed what happened in this field.

In 2006, I was contacted by Nick Marsh, head of the Metropolitan Police photographic service, with whom I had served in Kosovo. He had a problem case he didn't quite know where to go with, and thought he would see if I could help. The Met were investigating a father accused by his teenage daughter of sexual abuse. They had some images they felt could be useful but they didn't know how they might extract evidence from them—and, frankly, neither did we.

The girl alleged that her father would come into her bedroom in the middle of the night and touch her inappropriately as she slept. She told her mother, who was not prepared to believe her and dismissed her complaint as attention-seeking. But this smart, brave young lady, determined to demonstrate that she was speaking the truth, set up the camera on her computer to record through the night. At 4.30 in the morning the camera captured images of the right hand and forearm of an adult male who entered the room and proceeded to interfere with her as she lay in her bed—just as she had said.

In the dark, the camera had switched to infrared (IR) mode and so the picture was in black and white. When the image of a live body part is recorded in this way, the near-IR light is absorbed by the deoxygenated blood in the superficial veins. As a result, the intruder's veins were clearly delineated, looking a bit like a map of black tram lines. The question we were asked was: could we identify someone from the pattern of veins on the back of their hand and their forearm? Absolutely no idea was the answer, but we would think about it and do some literature research to see what was already out there.

The number of publications relating to human anatomical variation is extensive. As well as having great importance for the medical, surgical and dental worlds, it has translatable value into the forensic realm of human identification. Andreas

Vesalius knew in 1543 that the veins in the extremities of the body were highly variable in their location and pattern and that, in terms of finding a vein in the upper limb where you expected it to be, nothing between the antecubital fossa (the pit of the elbow) and the fingertips could be relied upon with any great certainty. Some 350 years later, at the dawn of the twentieth century, another professor of forensic medicine at Padua University, Arrigo Tamassia, offered the opinion that no two vein patterns seen on the back of the hand were identical in any two individuals.

Tamassia was critical of the Bertillon system of anthropometry that was starting to gather momentum as a means of recording the physical measurements and appearances of criminals. Bertillonage, in tandem with fingerprinting, predominated in the scientific criminalistics world at that time. Given that vein patterns could not be disguised, did not change with age and could not be destroyed, Tamassia considered there to be a place for vein-pattern matching in the identification of offenders. On the basis that, while fingerprint analysis required lengthy training, the six basic patterns of vein analysis, and their many subclass variations, could be observed from a photograph or drawn on to paper, he argued that examination of the veins would be an easier test for law-enforcement officers to perform.

Tamassia's new technique was quickly picked up in the United States. Newspaper articles in the *Victoria Colonist* in 1909, and in *The New York Times* and *Scientific American* the following year, hailed it as revolutionary.

Tamassia rather sweepingly described vein patterns as 'infallible, indestructible and ineffaceable'. Perhaps this was a little rash, as his pronouncement quickly became immortalised in crime fiction by Arthur B. Reeve, author of a series of detective stories featuring Professor Craig Kennedy, who was dubbed by some the American Sherlock Holmes. In '*The Poi-*

soned Pen' (1911), Kennedy confronts the criminal with the words: 'You perhaps are not acquainted with the fact, but the markings of the veins in the back of the hand are peculiar to each individual—as infallible, indestructible, and ineffaceable as finger prints or the shape of the ear.'

Somehow science fell out of love with vein-patterning and its shine faded. However, like all good ideas, it did not die but merely lay dormant, waiting for opportunity and happenstance to bring it out of hibernation. In the early 1980s Joe Rice, an automation controls engineer at Kodak in England, believed he had invented vein-pattern recognition. He hadn't, of course, because Vesalius and Tamassia had blazed that trail. What Rice did invent, using infrared technology, was a biometric vascular barcode-reader which could store his own vein pattern and, by extension, those of others. He came up with the idea after his bank cards and identity were stolen, and developed it into a method of identification that he claimed was more secure than a PIN number.

Rice patented his Veincheck system, but the world was still wedded to fingerprints and his innovation, like Tamassia's before it, fell into relative disuse. By the new millennium, however, biometrics and security were a boom industry. With the expiry of the patent on Rice's work, both Hitachi and Fujitsu launched security products using vein biometrics, hailing vein patterns as the most consistent, discriminatory and accurate of biometric traits. Today security experts view vascular pattern recognition (VPR) as a valuable biometric because, they say, it cannot be destroyed, it cannot be mimicked and it does not change with age. Isn't it wonderful how history hums a familiar tune?

For the vein pattern to be used as a means of identification it must first be recorded and stored in a searchable database. When the individual then places his or her hand on an IR viewer, the image is compared automatically with all the patterns held on the database and is matched to its owner. There is no risk to health and, since the hands are always on show, there is no stigma or inconvenience attached to presenting this part of our body for examination.

To see the variability of the vein patterns in the human body, just take a look at the veins on the back of your left hand, compare the pattern to that on your right hand and then go and check them against someone else's. If your hands are a bit hairy or chubby, you may be able to see the patterns on the inside of your wrists more clearly. They will all be different, even in identical twins, because our veins are formed before birth in a way that makes them unique to us. In the fetus, blood vessels develop from small isolated 'puddles' of blood cells. When the heart starts to pump and then relax, the arteries and veins begin to assemble as the puddles co-alesce. Arteries are fairly consistent in their location and patterning; veins are much more variable, and the further away they are from the heart, the more variable they will be. That is why, as Vesalius observed, the veins in the feet and hands will show more pattern variation than those in the legs or arms.

By 2006, we had the benefit of the accumulated work of Vesalius (from anatomical specimens), Tamassia (from forensic research) and the advances made by Rice, Hitachi and Fujitsu in biometrics. Could we translate it all into a technique that would answer the question posed by the police investigating this particular accusation of sexual abuse?

What we did not have was the opportunity to compare a vein pattern extracted from an image via a mathematical algorithm with its counterpart stored on a database. We had to compare the image from the young girl's computer camera with a custody suite photograph of her father's forearm and hand. So in that sense our method was more akin to Tamassia's than to those of his successors. If the vein pattern was not the same, we could be certain that the arm and hand could not belong to the same individual in both images, and therefore we would be able to exclude the father. If the pattern was the same, we could not, however, say with equal certainty that it was the same man in both images because we had no information that would allow us to make the necessary statistical inferences as to exactly how variable a vein pattern would be, and whether two people might share the same pattern. We could hardly ask Vesalius, who had been dead for 500 years, or Tamassia, who had been gone for nearly a hundred, so we were fundamentally looking to be able to exclude the father. I wonder whether Vesalius or Tamassia could have given us an answer. At times I feel that collectively we may have forgotten more anatomy than we remember.

It is important in forensic science that we do not overstate the capabilities of any technique. It is not our job to find someone innocent or guilty, it is our duty and responsibility to examine the evidence dispassionately and give a valid professional opinion, honestly and transparently expressed, on the reliability, accuracy and repeatability of our methods and our findings.

Having compared the veins on the offender's right hand and forearm with those of the girl's biological father, we went to court with our results and opinions, such as they were. Since this was the first time that such evidence had been heard in a UK court, there was a lot of discussion between the judge and the legal teams about its admissibility. The jury were asked to retire from the courtroom to allow voir dire—a preliminary examination of a witness by the judge or counsel to assess the admissibility of the evidence. The judge ultimately decided that, because vein-pattern analysis was based on sound anatomical knowledge and a prior, albeit limited, research history in the biometrics industry, it would be admitted into the court and so the trial went ahead and we presented our evidence. There was an understandably robust cross-examination from the defence team, but it was survivable.

When the jury returned a not guilty verdict we were more than a little surprised. What were the chances of someone else—and someone with a very similar superficial vein pattern on their hand and forearm—being in the teenager's room at half-past four in the morning? But it is not our place as expert witnesses to convince a jury or question their findings: they are the triers of fact and the final decision is entirely theirs, on the direction of the judge.

What we could do, and did, was ask the barrister if she thought there had been any problem with the science, or my communication of the science. Perhaps I had not been able to convey the information clearly to the jury? It transpired, somewhat bizarrely, that the barrister felt my evidence had probably not been a critical factor in their ultimate decision. Her impression had been that the jury simply had not believed the child. It was possible that they thought she was not sufficiently upset and that maybe her demeanour sowed doubts in their minds that she was telling the truth. And so the defendant was free to continue living, as an innocent man, in the very house where his own child had accused him of abuse.

What became of the girl I will never know, but it worries me to this day that perhaps I could have done more. There is only one way to improve the quality of the evidence we can provide and its prominence in a case, and that is to increase the validity and robustness of the science, which was what we now set out to do. There was clearly some merit in Tamassia's original research and we wanted to bring it into the modern world of the analysis of indecent images of children (IIOC).

As well as being a barbaric betrayal of the trust our children place in adults, the taking and sharing of IIOC is one of the fastest-rising crimes of this millennium. We decided to follow in the footsteps of Vesalius and Tamassia by initially studying variation in the venous anatomy of the back of the human hand. This is the part of the perpetrator's body that is most frequently captured in such images. It was serendipitous that between 2007 and 2009, the DVI training courses Dundee University was providing for police forces across the UK would bring over 550 officers to our campus. We asked if they were willing to assist us with the creation of a database so that we could investigate anatomical variation and virtually all of them agreed.

We were not looking only at veins: we also started to consider scars and the patterns of moles, freckles and skin creasing at the knuckles—all regarded as 'soft' biometrics. We discovered that, when analysed in combination, these independent variables were extremely useful in assisting us to distinguish reliably between individuals. We photographed every officer in IR and visible light, recording their hands, forearms, arms, feet, legs and thighs. The result is a unique ground truth database which has proved to be of great value in the validation of our research.

We got the grants, we did the research, we wrote the papers and we have now helped the police with over a hundred cases of alleged child sexual abuse, in which we have been able to help to exclude those wrongly accused and to supply evidence for the prosecution of the guilty. We have worked with most UK police forces, many across Europe and some as far away as Australia and the US. By the time a case comes to us, the police generally have a fairly strong idea of who the perpetrator might be and there is usually a significant amount of evidence to support the Crown case. However, in many instances the accused will either plead not guilty or, on the advice of their legal counsel, will present a 'no comment' response throughout the interview. In the cases that we have accepted, over 82 per cent of defendants have subsequently changed their plea to guilty as a result of the additional information our analysis has offered.

This is incredibly important because it means a full court trial is no longer required. Not only does it save large sums of public money, but even more importantly, it means that the victim does not have to give evidence in court against their abuser, who may be their father, their mother's boyfriend or someone else they know. It is enormously satisfying to have played a small part in cases where the courts have handed down sentences amounting to several hundred years of incarceration, including many life sentences, for those who commit what I consider to be one of the most despicable and heinous of crimes against the most vulnerable in our society. No adult has a right to steal the innocence of childhood.

This success has been achieved in no small part thanks to anatomy, where the dead really do continue to teach the living—not only by lending us their bodies but, in the case of the lessons of Vesalius and Tamassia, through the legacy of their work.

CHAPTER 13

An ideal solution

'I profess to learn and to teach anatomy not from books but from dissections, not from the tenets of Philosophers but from the fabric of Nature'

William Harvey, physician, *De Motu Cordis* (1628)

A cartoon donated for our 'Million for a Morgue' campaign.

The dead must have somewhere to reside until they move on to their final resting place and those who bequeath their remains to our anatomy department have chosen a calm waiting room full of people who care. To demonstrate their belief in our job, many staff in anatomy departments sign bequeathal forms so that when their time comes—we hope after a long and happy retirement—they will return to their workplaces to resume their role as teachers. It is somehow in keeping with the discipline that their life's work also becomes their death's work.

The trade we ply in this place of death may seem macabre to some but in reality it is anything but. Our donors often have the most wonderful sense of humour. The elderly gentleman tickled by the fact that 'a young lassie like you still wants my old body' was not untypical. And many feel quite strongly that their remains should be put to worthwhile use. Allow me to share the words of Tessa Dunlop, who wrote to us about her father, a no-nonsense Perthshire farmer.

> My father, Donald, has been terminally ill with a bone marrow cancer for more than four years—his once enormous frame is now ravaged. I couldn't imagine that science would have much use for him. In fact, I wasn't at all sure that science still needed bodies. No one ever seemed to talk about a shortage. Hadn't computers replaced the role of the cadaver? But Dad was adamant. 'A dead body is a very unattractive thing. You won't want mine hanging around, and I can't stand funerals. Surely a medical school will take me?' A couple of

> forms, a witness signature and a week later he got the answer he was hoping for. The University of Dundee . . . accepted his 'generous offer'. He was grinning from ear to ear.

Mr Dunlop had worked hard all his life and to him it seemed only fitting that he should carry on working after his death. But while our donors might be fully at ease with their decision, sometimes it is not so easy for those left behind. A widow whose husband had bequeathed his body to my department once pleaded with me to 'look after' him as she did not fully understand his decision or his wishes.

To hand over the remains of the person you have loved all your life to a stranger must be a hard thing to do and we take our duty of care very seriously. Indeed, since human anatomy and the world of the corpse are at the core of everything we do, it is our priority every single day to guard and protect zealously those who have chosen to continue to be of value to society after their death by helping us all to learn more about the human body. That widow must have been won over by her experience of our department, because by the time of her husband's funeral service she, too, had signed her forms to donate her body to anatomy. This is something that, encouragingly, happens quite often.

When I moved to the University of Dundee it was on the guarantee that full body dissection would be retained for all our students. In the eyes of most university deans, anatomy is a dead subject from which there is no further financial return to be gained and is therefore an expensive luxury. As a result, it has become a focus for disinvestment in many medical schools. Administrators may be seduced by the augmented or virtual reality options modern technology can offer, which they believe suit the short attention span associated with education today. But to assume that there is nothing new to learn, and no operational processes to be amended or developed, in the study of anatomy is to misunderstand its importance to so many disciplines. In a lazy world, it is much less trouble to pronounce a subject dead than to look at what is required to rejuvenate and expand it.

No computer, book, model or simulation could ever replace the multisensory impact of learning from the gold standard. To go down the road, as so many anatomy departments have done, of depriving students of the opportunity to explore a real human body is, in my opinion, detrimental to their university experience and simply sets up problems for the doctors, dentists and scientists of the future who, to become experts in their field, surely deserve to be able to study through the best medium. The one thing every student learns in a dissecting room is that no two bodies have identical anatomy. There are so many possible variations and if these are not learned and understood, the people who will suffer are the unsuspecting patients of these future practitioners. Approximately 10 per cent of all legal cases involving surgical malpractice are believed to be due to ignorance of anatomical variation.

Anatomy has been governed by an Act of Parliament since 1832. The first legislation was introduced principally in a knee-jerk reaction by Earl Grey's Whig government to the murders committed by Burke and Hare in the West Port district of Edinburgh. In an attempt to halt the illegal trade in corpses and to raise the ethical standards of the profession, the law permitted teachers of anatomy to dissect the bodies of both the criminal and the unclaimed poor and to accept bequeathals and donations.

Before recent amendments to the Anatomy Act—in 2004 in England and 2006 in Scotland—an old clause persisted which, paradoxically, made it illegal for surgeons to practise or test procedures on a dead body. They could come into a dissecting room, they could cut a cadaver's skin, move the muscle or saw out the femur, but they could not replace the bone with a prosthesis, as that was categorised as a 'procedure'. This restriction was a long-lasting and pointed reminder of the historical relationship between surgeons and anatomists. Good old Burke and Hare still have a great deal to answer for.

Many anatomists, surgeons and clinical practitioners gave evidence to government committees to reassure them that the nefarious commercial association behind this 170-year-old ban was no longer relevant—surgeons could be trusted and should be allowed to hone their skills on the human cadaver rather than on the hapless patient. The age-old partnership between surgery and anatomy began to blossom once again but there was a minor hurdle to be overcome first. Surgeons turned their backs on anatomy very soon after the legislation was changed because they found that the formalin we used to embalm our cadavers rendered the body too stiff and inflexible for their purposes. They wanted something that more closely resembled the feel and tissue response of a live patient and decided that they would instead prefer to pursue the option of using 'fresh/frozen' cadavers.

I had more than a passing objection to anatomy being driven in this direction. Let me explain. To comply with the surgeons' preferred approach, when a bequeathed body is received fresh from the hospice, hospital or wherever the donor has died, it needs to be sectioned into relevant pieces (shoulder region, head, limbs and so on) in a manner fairly reminiscent of dismemberment, which criminal law views as an aggravated insult to a corpse. The parts are then frozen and, when required for surgical courses, recovered from the freezer, defrosted and worked on by the students, surgeons in training and other groups. Because, once defrosted, these sections are to all intents and purposes 'fresh', after a couple of days they will no longer smell quite so fresh and, like any organic matter, will not respond well to repeated freezing and thawing. They will therefore be of limited, if any, value for others to study after the first group has finished with them. Furthermore, many pathogens are known to be able to survive freezing and can merely lie dormant before regenerating when the tissues begin to warm up again. The likelihood of transfer and infection then rises and great care must be taken by those learning their trade not to cut themselves as well as to ensure that all their inoculations are up to date.

The change in the law also made it legal for us to import body parts into the UK from overseas. Now, this made me very nervous. The fact that you could place an order with a company in America for, say, eight legs, all carrying an alleged clean bill of health, was bad enough; the fact that after use these body parts would then be incinerated as clinical waste I personally found disrespectful and unacceptable. It felt to me as if these remains were being treated more as disposable commodities than with the consideration due to people who had died.

So for me, the fresh/frozen system as administered in some institutions appeared not only wasteful of a precious resource but also morally questionable. Neither was I prepared to take the inherent health-and-safety risk. I could imagine the possible consequences: if a surgeon or a student cut themselves and contracted an infection, we could be facing a lawsuit claiming that the injury had ended their medical career. Our university undertook a feasibility study, led by a member of the senior management, and I was delighted and relieved when this produced the recommendation that Dundee should not adopt the fresh/frozen method. If it had, I would have left.

It was clear, though, that continuing with formalin was a problem for a number of reasons. For a start, there was the cost. We had seen from the experience of the trainee surgeons that human tissue embalmed in formalin was not ideally suited to all the procedures they needed to practise. And the critical issue of health and safety was paramount to medical and reputational security. It is important that tissue should be sterile, and while formalin does satisfy this criterion to a greater or lesser extent, we also knew that, in concentration, it is a potential carcinogen. Indeed, formalin was already coming under some scrutiny in many countries and its use was beginning to be discouraged in the wake of a 2007 EU ruling under which a reduction in the approved concentration was being considered. If further dilution became a legal requirement, formalin's role in the field of anatomy could become redundant. We would have to get our clever heads on and find a solution that would meet everyone's needs.

I remembered hearing about a technique being used in Austria, developed by a charismatic and inspirational anatomist and teacher named Walter Thiel, and started to wonder whether maybe this could provide an answer. Professor Thiel, head of the Graz Anatomical Institute, had been a medical student in Prague when he was called up for active service in the Second World War. After being invalided out of the army with a gunshot wound to his face, he overcame his injuries and resumed his medical studies. After the war, he dedicated fifty years of his professional career to the Graz institute where, in the early 1960s, he recognised the very problem we were wrestling with and made it his life's work to solve it.

Thiel's goal was to find a better way of preserving bodies, one that would leave the tissues flexible without compromising the longevity of the material, and which would at the same time ensure a healthier working environment for anatomists and students. He had noticed that the quality of the texture of the 'wet-cured' ham in his local butcher's shop was infinitely superior to the end product he was able to achieve in his embalming rooms. The ham retained both its colour and its flexibility after being preserved in a solution of salts. It occured to him that perhaps the food industry may have something new to teach anatomy.

The butcher was of course restricted to using chemicals

suitable for the food chain that would not poison his customers. Walter Thiel had no such limitations. His medium was never going to be sold over the counter for consumption. So he embarked on a painstaking process of trial and error to perfect a soft-fix solution—pickling, basically—using a combination of water, alcohol, ammonium and potassium nitrate salts (to fix the tissues), boric acid (for its antiseptic properties), ethylene glycol (to increase plasticity) and just enough formalin to act as a fungicide.

Beginning his trials with cuts of beef from the same butcher's shop, he worked his way up to whole animals. He realised that it wasn't sufficient just to perfuse the body, it also had to be immersed in the fluid for quite a long time to ensure that it was preserved from the inside out and the outside in. The tissues then remained fixed, retained colour and plasticity and did not need to be refrigerated. Importantly, there was also no sign of bacteria, fungus or other pathogens. It took him thirty years and over 1,000 bodies to arrive at a formula with which he was finally satisfied, and which he felt would be effective for a period that was optimal for the dissection of all the tissues. The last and best solution he manufactured was virtually sterile, virtually colourless and almost odourless. It did everything he required of it, and it was relatively cheap to produce.

The motto by which Walter Thiel lived—'Only the best is just good enough'—his infectious optimism and his unbreakable spirit were all reflected in his determination to make a difference in his chosen subject. Probably much to the chagrin of his university today, he did not patent his method. It is testament to his generosity of spirit and firm commitment to collaborative research and learning that he instead published it to the world, believing that such scientific developments should be open to all and not appropriated to make a profit or for the advantage of one school over another. His ethos resonated loudly with us.

It all sounded too good to be true. But sometimes it isn't necessary to go back to the drawing board yourself. Someone else may have got there before you, and all you have to do is tailor their work to your own requirements and develop it further—just as we had done with Tamassia's vein-identification hypothesis. We don't all need to be geniuses; some of us just need to be practical applicators and adapters. It is important that we are honest and do not take credit for ideas that did not originate with us.

I sent two members of my staff, Roger and Roos, over to Graz on a reconnaissance mission to check out the Thiel technique. They came back absolutely bouncing with enthusiasm for the possibilities it offered. They talked about the incredible flexibility of the cadavers, the duration of their preservation, the welcome absence of the stench of formalin that pervaded every UK anatomy department and the apparent resistance of the embalming fluid to any propensity to culture bacteria, mould or fungus. They had found no obvious downside. All this and it was, they said, no more expensive than formalin . . . except in one teeny, tiny, little area. It would require an entirely different mortuary set-up, and that meant capital costs: the very type of funding universities do not want to allocate to what they perceive as dead and dying anatomy departments.

That was a bridge we would have to cross when we came to it. First of all, we persuaded Sir Alan Langlands, the principal of our university, to give us a small amount of money, about £3,000, for research and development because we needed evidence from a proof of concept. We decided to trial the Thiel embalming on two cadavers, male and female, to whom we referred in-house as Henry and Flora. (Why do I always seem to name my male cadavers Henry? Maybe *Gray's Anatomy* is so ingrained in my soul that it is subliminal.) As we didn't have the necessary custom-built facilities, we would have to create a suitable mortuary set-up for the trial. Born into a generation that was glued to their television sets for every episode of *Blue Peter* and *Dad's Army*, I pride myself on being able to devise a way to make almost anything out of sticky-backed plastic, washing-up-liquid bottles and cardboard toilet-roll tubes. What we couldn't make, we would have to improvise, beg and borrow (but never steal).

We found an old giant fish tank that was being thrown out of a decommissioned zoology building and turned it into a body-sized submersion tank that would hold two cadavers cosily side by side. We borrowed tubes and pumps. We fashioned lids out of old doors and got our heads round some pretty basic bucket chemistry. We also had to persuade the local police that the nitrates we were buying in bulk were not for constructing bombs, just for embalming cadavers.

Henry was the first to be prepared for Thiel embalming. The fluid we had made was pumped gently through the veins in his groin region and a small cut was made on the top of his head to get into the venous sinuses, which drain blood from the brain. The whole process took less than an hour. He was then submersed in our tank, where he was joined by Flora a few days later. There they lay for two months. We checked them every day, turning them to make sure that all surfaces were exposed to the fluid. We looked for signs of decomposition or bloating. Nothing seemed to be happening that gave us any cause for alarm. What we did notice was that every time we turned them over, they remained flexible and hard to catch, a bit like wet fish in a barrel. This was an encouraging sign.

Gradually the pink colour of the skin paled and the top layer of dead skin sloughed away, as did all the hair and nails. Surprisingly, as the skin swelled a little, wrinkles started to disappear and Henry and Flora began to look younger. Sadly, though, I don't think we discovered a new elixir of youth—the chemicals are too dangerous and lying in a vat of them for two months might be a bit inconvenient. As the weeks went by, nothing seemed to be going wrong. But we spent a lot of that time with our fingers and toes crossed.

We wrote to all our surgeons in the Dundee and Tayside area and asked them if they would be prepared to trial various surgical procedures using the Thiel approach and complete a feedback evaluation on what worked and what did not work from their point of view. They were generous with their time and advice. Roger and Roos planned the procedures like a military operation to make sure that the least intrusive were completed first and the most intrusive last, in order to make the most of Henry and Flora's pioneering assistance. Every surgeon reported that the Thiel-preserved tissue was far superior to that embalmed in formalin and much more pleasant to work with than fresh/frozen, while offering the same advantages. Indeed, from their perspective, the only difference between the Thiel cadaver and an actual patient was that the bodies were cold and there was no pulse. Is this another gauntlet I see before me?

While there isn't much we can do to raise the temperature of a cadaver, we have since had a go at giving them a partial pulse for one of our surgery courses. If we restrict an area of the arterial system, we can fill it with a fluid of the same consistency as blood and hook it up to a cyclical pump, producing what is technically a pulsatile blood flow. We can then simulate a haemorrhage and set a timer to count down the number of seconds the surgeon would have in a real-life situation to stem the flow before the patient would be lost. It is just the most amazing learning experience, with direct and immediate significance for patient survival and surgical skills, especially in combat surgery, where time is absolutely of the essence. We also found that we were able to hook the cadavers up to a ventilator so that we could simulate breathing. This makes 'operating' much more realistic but I have to admit that even I was a little bit unnerved at seeing a cadaver in my dissecting room 'breathe' for the very first time.

The entire project was a resounding success. We wrote to Walter Thiel, who was becoming quite frail by that time, to tell him that we had managed to emulate his fantastic achievements. Once again, we had developed nothing new: we had just listened to some old records and picked up the tunes. I was invited back to a meeting of the University Court to report on the outcomes of our trial. There was unanimous agreement that we should start planning for a full conversion from formalin to Thiel as soon as was practically possible. We saw no value in half measures and the university recognised the worth of becoming home to the only Thiel cadaveric facility in the country. It would make Dundee UK leaders in the field.

Once the decision had been made, we hit them with our tiny snag: our existing mortuary was not fit for purpose and it was too small. It barely met our current needs, let alone what we had planned for the future. As we couldn't possibly cease body intake while we renovated it, we would have to build a new one. In the meantime, Her Majesty's inspector for anatomy entered the fray. Our current embalming facility, the inspector said, was in desperate need of refurbishment and the university would need to consider updating it as a priority if they wished to continue to teach dissection-based anatomy at Dundee. This slightly orchestrated pincer movement was a bit naughty, I know, but the university undoubtedly had some decisions to make and this concentrated their collective mind. Should they maintain dissection-based anatomy teaching at Dundee? If so, should they refurbish the old mortuary and stick to formalin like everyone else, or should they fasten their Superman cloak around their neck, pull their underpants over their tights and make the giant leap of investing in a clear opportunity to claim a leading role in anatomy in the UK? Naturally, I happen to think that the decision they made was the right one.

◊

The new build would come with a hefty price ticket of £2 million. The university was prepared to meet half the funds but it would be up to us to raise the balance. Where on earth do you find money to fund a mortuary? A bag-pack at a supermarket or rattling a tin at the local railway station wasn't going to cut it and, let's face it, without a carefully thought-out campaign, an appeal for a new mortuary was unlikely to tug at the heartstrings of our generous and compassionate public in the same way as many of the charitable causes with which we would have to compete. We would simply have to get creative again and think way outside everybody's box.

I consulted a good friend of mine, Claire Leckie, who is a very experienced charity fundraiser. She suggested I compile a list of all the people who had 'used' my services over the years, because maybe it was time to call in a few long overdue favours. Who had, for example, picked my brains in the past, and could they help me now? When I started jotting down names I realised it was quite a long list, but one leaped out: the celebrated crime writer Val McDermid.

Val and I had 'met' on a radio programme over ten years before, when she was in Manchester and I was in the Aberdeen studio. While waiting to go on air we had been chatting and, on the spur of the moment, as is my wont, I had said to her, 'By the way, if you ever need any forensic-science advice, feel free to give me a call.' She did indeed call, many times, and it led to a warm and genuine friendship of which I am deeply proud. I knew that if anyone was brave enough, and bonkers enough, to help me with this, it would be Val.

We put our heads together to hatch a scheme and eventually came up with the wonderful 'Million for a Morgue' campaign. The new mortuary would need a name, and we agreed that it should be one that would resonate with the public and the media. But while an Olympic cyclist or an artist would consider it an honour to have a velodrome or an art gallery called after them, who would want to be linked with a mortuary? The answer was staring us in the face. The perfect fit would be a crime writer. And why not use the naming of the building to raise the profile of the campaign, and boost its coffers, by inviting the public to choose the recipient of this dubious honour?

Val persuaded a number of her big-hearted crime-writing colleagues to join her in supporting us, including Stuart MacBride, Jeffery Deaver, Tess Gerritsen, Lee Child, Jeff Lindsay, Peter James, Kathy Reichs, Mark Billingham and Harlan Coben. We launched an online competition in which crime-writing enthusiasts could vote for the mortuary to bear the name of their favourite author in return for a small donation. We all had a tremendous amount of fun and the writers' generosity and ingenuity knew no bounds.

Jeffery Deaver exhorted readers to vote for him on the basis that he was the one who looked most like a cadaver. I could not possibly comment, but nobody disputed his claim. A talented musician, he donated a CD of his private work to be auctioned to raise funds. Our only worry was what we would do if Lee Child won the most votes, as christening the new building the 'Child Mortuary' would hardly be sending out the right message. Lee, ever the gentleman, came to our rescue by suggesting that if he won we could call it the Jack Reacher Mortuary instead, after his most famous character. I have not yet exploited the connection to Tom Cruise, but I have tucked that away for future use.

The magnificently creative Caro Ramsay put together a 'killer cookbook' featuring recipes contributed by her fellow crime writers, which was sold in aid of the mortuary. Never before had a cookbook been associated with an anatomy department—for several very good reasons. That took a little bit of careful advertising, I can tell you, but it was a great success. We held tasting nights and cookery demonstrations all round the country and it was even shortlisted in the 2013 World Cookbook awards.

Other authors put characters in their forthcoming books up for auction. Winning bidders could choose to become a bartender, say, or an innocent bystander in their favourite crime writer's next novel. Stuart MacBride hosted tours around the Aberdeen locations immortalised in his Logan McRae books. Stuart also donated to the campaign the proceeds of *The Completely Wholesome Adventures of Skeleton Bob*, three children's stories he had originally written and illustrated for his nephew, Logan, recounting the adventures of a skeleton with a pink knitted skin who gets into all sorts of trouble with witches and his father, the Grim Reaper. We were touched and honoured that he entrusted them to us and allowed us to publish them on his behalf.

For a full eighteen months, we worked our socks off for the Million for a Morgue campaign. We did signings at crime-writing festivals from Harrogate to Stirling. We gave lectures and interviews, we talked about our mission on television and in newspaper and magazine articles. We arranged and participated in discussion panels. And we did it. We raised what was needed to fund the shortfall and to enable construction of the new Thiel facility to begin.

Focused as we were on building and equipping our new mortuary, we were completely taken aback by one unexpected side-effect of our campaign. Back in Dundee, Viv, our bequeathal manager, complained that after every event in which I took part there would be a surge in bequeathal inquiries from people who, until hearing about Million for a Morgue, had never realised that anatomy departments still required bodies for teaching and research. They went on to sign the donation forms in their droves, not just to Dundee, but to other anatomy departments across the UK.

We had never envisaged our campaign as a kind of a

recruitment drive for the dissecting room, but it seemed that this was proving to be a very positive and welcome by-product. Indeed, long after the end of the fundraising activities, the interest still continues. Today we have more than a hundred donations annually, almost all from the Dundee and Tayside region, with which we have built such a strong bond of trust.

We hadn't been appealing for anything other than money to help us build the new facility, but perhaps we should have been. Why do we always think the public needs to be wrapped in cotton wool when it comes to death and that people won't want to talk about it? All those who contacted us were relieved to find that there was a matter-of-fact discussion to be had and a third option for them to consider, as well as burial or cremation, when the time came to decide what should happen to their remains after they died. They showed no reluctance to talk about death or to ask direct questions, and no difficulty dealing with direct answers.

One lovely lady seeking to donate her body to Dundee was calling from Brighton on the south coast of England. Viv, as professional etiquette demands, quite rightly pointed out to her that there were medical schools much closer to her than ours. We could, of course, accept her donation but her estate would have to meet the transport costs. She said she didn't care about that. She wasn't interested in going to a local medical school because she wanted to be a state-of-the-art cadaver—she wanted to be 'Thieled'. Walter, who passed away in 2012 and therefore sadly didn't live to see the opening of our new mortuary, would have been so proud and endlessly amused to learn that he has become a verb.

At one of the fundraising dinners we held with the crime writers we met a very conflicted woman. She was terminally ill and determined to donate her body to anatomy but her husband was very much against it. While she did not want to upset him, she did want him to respect her wishes and to agree to fulfil them. During our lengthy conversation, it became clear that her inability to make her husband see how much this meant to her was a source of some distress. He was understandably afraid that we did unspeakably disrespectful things to cadavers and his sole concern was the responsibility he felt to protect her dignity and decency in death. She asked if I would be prepared to write to her explaining exactly what we do and why we do it in the hope that such a letter could be used as the starting point for a less fearful marital discussion.

It was a hard letter to write and it took me a long time, but it had its reward in her response. She said her husband thought he 'got it now' and that, although he still 'wisnae happy', he had agreed to respect her decision. I can only hope that she got her last wish, that her body resided for a while in an anatomy department somewhere in the central belt of Scotland and that her husband has been able to draw some small comfort from knowing that what she has taught a generation of students will go on to benefit the sick and dying for longer than she could ever have imagined.

Whether our donors come to us in Dundee, or we can help smooth their path to another anatomy department to fulfil their final request, it is an honour and a privilege to assist them with their plans for 'being dead'. Whatever their job or station in life, whether they are rich or poor, tall or overweight, riddled with disease or arrive with a full manicure and new hairdo, whether they die too young or at a great age, these amazing people are united in their decision to donate their remains for the common good of priceless education.

We consider it our duty as licensed teachers of anatomy to speak on their behalf, to stand up for the principles they represent and to preserve their dignity. Mercifully, long gone are the days of comic films in which routines involving cadavers bundled into taxis, or fingers found in someone's breakfast, portrayed the dead as a prop in the capers and high jinks of disrespectful medical students. I tolerate no disrespect in my dissecting room, and neither will Her Majesty's inspector for anatomy. Contravening the Anatomy Act can carry a custodial sentence. Quite rightly, given how much faith and trust our donors place in us and our students to do our jobs.

It is this sense of responsibility that underpins my personal views on the public display of cadavers and the tipping point at which this can no longer be justified as educational and becomes nothing more than ghoulish voyeurism. Charging a high admittance fee in the name of education to bring in the public to gawp at cadavers posed as if they were playing chess or riding a bike, or vulnerably exposed in their third trimester of pregnancy, does not make such an exhibition educational. I find the showmanship element distasteful and cannot think of any circumstances in which I would ever support such a commercial venture. With the permission of our inspector, we have been allowed, from time to time, to place embalmed specimens—the ones we have mounted in glass or Perspex pots—in special exhibitions at science centres and other such venues where there are no entrance fees and where the focus is truly on education. But we can't, and won't, do it just for the sake of entertainment. There has to be a distinct educational purpose to it, and this will always make fundraising an uphill struggle for anatomy.

For us, our subject is very far from being either dead or dying. It is alive and kicking in pockets around the world where its adherents remain passionately committed to its survival and its growth—and nowhere more so than in Dundee. I am so proud of all the donors, the staff, students and the many supporters who have made this phenomenal educational and research facility one of the best in the world. In 2013, we were honoured to be awarded the rare and prestigious Queen's Anniversary Prize for Excellence in Higher and Further Education for our research in human anatomy and forensic anthropology.

And the forward thinking goes on. At the time of writing, we were planning to mark 130 years of teaching anatomy at Dundee in 2018, and gearing ourselves up for another campaign to raise funds for adding a new public engagement centre to our building.

And what of our brand-new mortuary? That was formally opened in 2014 as, to the surprise of nobody, the Val McDermid Mortuary. There was never much doubt, given her massive following and in recognition of her huge drive and commitment to our cause, that Val would win the competition. Because of his enormously significant contributions—make no bones about it, little Skeleton Bob unquestionably left his mark—and because he gained the second-highest number of public votes, we named our dissecting room after Stuart MacBride.

In acknowledgement of the generosity of the other writers who lent us their reputations, time and effort during the campaign, we decided to name individual Thiel submersion tanks after nine of the key players. The tenth is dedicated to my former principal anatomist, Roger Soames, who was a stalwart support throughout, and indeed throughout everything we have done at Dundee. He retired shortly after the mortuary was built, so we christened a Thiel tank after him as a farewell gift. When people see his name on the tank, they assume it is him in there. It is not. Roger is happily and healthily retired, but who knows, maybe one day my favourite anatomist and dearest friend will come home to teach his students again. If he does, he will be welcome, but I hope it will be at a very distant future date.

We have eleven tanks altogether. I quite fancy the idea, when my time comes, of floating peacefully in the Black Tank. How cool would that be?

Epilogue

'To die will be an awfully big adventure'

J.M. Barrie, *Peter Pan*

Life's contemplation.

Through this brief exploration of the many faces death has revealed to me, I hope it will have emerged that my relationship with her is one of comfortable camaraderie.

Although I am no scholar of thanatology—the scientific study of death—I think I have experienced enough of her handiwork to have gained a healthy understanding of what might be coming my way. I would, however, never be so bold as to predict with certainty how I might behave at the end of my life. I suspect that someone who muses deeply and often on his or her dying and death bears little resemblance to the person who finally confronts her in the raw. It is the element of the unknown that prompts such philosophising, which tends to increase as the years pass and the edge of our own hole in the ground looms ever nearer. Since nobody has ever returned to tell us what death is really like, no amount of preparation and planning can guarantee the smoothness of the path that lies ahead for us. The only certainty is that we will all have to walk it sooner or later. And although others may walk part of the way with us, it is a journey we must ultimately make with only death herself for company.

The point at which we feel living has stopped and dying has started will, I imagine, be different for all of us. For many, simply not being dead means we are not yet done with living. Is there anything we can do to hold her at bay? Perhaps she is open to some degree of mortal advocacy. Perhaps it is possible to reason or bargain with her, if the debate we place before her is sufficiently compelling and bolstered by a strong attitude of mind. How often have we heard of the terminally ill patient who, determined to see their last Christmas, their child's wedding or some other significant event, outlives a clinical prognosis to achieve their wish, only to die days afterwards? The trouble with a prognosis—which, after all, can only ever be a guess—is that it has a habit of becoming a self-fulfilling prophecy. Maybe it sometimes strips us of our will to keep fighting beyond the deadline we have set ourselves and thereafter we lose focus, just let go of living and start dying. Or maybe we have invested every ounce of strength we have left in reaching that milestone and are simply spent.

Summoning the will to battle encroaching death constantly and relentlessly, rather than focusing on a specific goal, may be another alternative. The real inspiration in this regard is Norman Cousins, the American political journalist who, on being diagnosed in 1964 with the crippling connective-tissue disease ankylosing spondylitis, was told that he had only a 1 in 500 chance of recovery. Having long believed that human emotions were the key to success in fighting illness, he started to take massive doses of vitamin C, moved into a hotel and bought a movie projector. He found that if he could get a really good belly laugh from watching reruns of *Candid Camera* episodes or Marx Brothers films, he would have at least two hours of pain-free sleep.

Within six months he was back on his feet, and in two years he had resumed his full-time job. Cousins died of heart failure twenty-six years after his diagnosis—and thirty-six years after first being told he had heart disease. He simply refused to die when the doctors told him he would and his therapy was laughter. There is nothing wrong with letting go of life if that is our choice, but maybe his experience is a lesson to those of us who are not yet ready to do so.

There are many well-known factors that can have either a beneficial or a detrimental effect on our longevity. A healthy diet, exercise, being married and being female are all likely to result in a longer life. The fact that women's lives will be about

5 per cent longer than those of men is borne out in virtually every country that has been studied. There is a suggestion that this may be because women have two X chromosomes and men only have one, which gives women a spare if something goes wrong. It's a nice idea, but the inferior longevity of the male is much more likely to be due to the negative side-effects of testosterone.

A research study on the eunuchs of the Imperial Court of the Chosun Dynasty (1392–1910) in Korea showed that they lived on average twenty years longer than non-castrated men. Interestingly, though, this was only true if their testes were removed before the age of fifteen. For individuals sterilised after the onset of puberty, and therefore after the biochemical influences of testosterone had kicked in, the differential was less striking. But it would be somewhat extreme, not to mention significant for the future of the human race, if men were to try to gain themselves another twenty years by getting themselves sterilised.

We usually measure our life, and its constituent parts, in weeks, months or years. It might be more interesting to measure it in risk. There are credits and debits here that can affect our life expectancy, and choosing how to play them can have some influence on the likely outcome.

In 1978, in a contribution to the book *Societal Risk Assessment: How Safe is Safe Enough*?, Ronald A. Howard of Stanford University introduced his concept of a unit of risk of death, which he quantified as 1 in 100,000 and named the 'micromort'. The principle is very straightforward: the higher the value in micromorts of a particular activity, the more dangerous it is and the greater the chance it will result in your demise. It can be applied to both daily tasks and more hazardous enterprises, and to those carrying immediate or cumulative perils. For example, 1 micromort equates to travelling six miles on a motorbike or 6,000 miles by train, the implication being that, as a mode of transport, a train is 1,000 times safer than a motorbike. So this measure allows us to compare the risk inherent in various activities and might, in some instances, make us think twice about whether a particular venture is really worth the gamble. An operation with a general anaesthetic is valued at approximately 10 micromorts, skydiving at around 8 micromorts per jump and running a marathon about 7 per run. The real risk-takers can rack up an impressive number of micromorts—mountaineers may expose themselves to 40,000 with each ascent.

These are all single acts carrying the danger of instant death, defined by Professor Howard as an acute risk. An activity with a cumulative effect, which will take time to become a genuine risk factor, is termed a chronic risk. In this category, drinking half a litre of wine or living for two months with a smoker will earn you 1 micromort.

On a happier note, we can buy back some of our endangered time by acquiring microlives for ourselves. The microlife is a unit quantified by Sir David Spiegelhalter, of Cambridge University, as a daily gain or loss of thirty minutes of our existence. We all know what kind of activities are going to earn or cost us microlives and, to be honest, those on the credit side are rarely fun. Four microlives for men and 3 for women looks very much like five servings of fruit and vegetables every day. Yep, raw cabbage for lunch again.

I think we should devise a new risk measure: the micromirth. How much more wonderful would our lives be, whether long or short, if we measured them in joy, laughter and utter nonsense? Microlives accumulate, micromorts are fatal but micromirths are priceless. I think Norman Cousins would agree.

◊

So what about my own dying, death and being dead?

I am quite relaxed at the moment about the 'death' and

'being dead' bits—they hold no fear and actually, I feel almost a slight frisson of excitement at the possibilities of what is to come. I have known the imperfections and strengths of this body all my life and I would really like to see how it copes with the task required of it before the final shutdown. I am no hero, though, so, in common with most people, I would sooner skip through the 'dying' part as quickly as possible. In an odd way I am quite intrigued by the threshold that separates dying from being dead, and I would like to experience that when the time comes. Just not for too long. As the Roman philosopher Seneca said: 'The wise man will live as long as he ought, not as long as he can.'

I have no desire to live to too great an age if that means being a drain on resources needed by younger people, especially if I have nothing of any value left to give and have become a burden to those I love. I want to be independent and mobile right up to my last hours on this earth and for that I would willingly sacrifice quantity for quality. Let me go out with a bang, not a whimper. I am prepared to tolerate some bodily discomfort with advancing age but please don't addle my mind. Don't let me languish in a soulless care home or hospital. Don't let dementia steal my life, my stories and my memories. I do not want my dying to echo my father's.

I have been asked why I decided to write this book, and why now. The truth is that it is an opportunity to set down some of my stories for our girls, so that they can always hear them in my words and not those of others. My father was a great storyteller and I listened to his tales time and time again as I was growing up. Recently I found a letter Grace and Anna sent him in 1997. As part of his Christmas present they had given him a book and a pen and asked him to write down his stories so that they would not be lost for ever. Sadly, he never did, and most of them died with him. A few more will eventually die with me. So I hope that this book will give Beth, Grace and Anna, and the generations to follow, a little more insight into me, and my life, when I am gone.

My husband and children despair of me because the last time I actively sought out a GP was when I was pregnant with Anna over twenty years ago. I take no prescription medication, although I suspect that if I went for an MOT, I would be put on a regime of tablets to alter my sugar, or my blood pressure, or my cholesterol or my something or other. Once you start down that road, you will be taking tablets for the rest of your life.

And the indignity of an invitation to a 'poop' test landing on your doormat on your fiftieth birthday . . . really! Of course I understand that preventative medicine saves lives, and there will be many who are glad they opted to undergo such tests. We all have a choice in these matters. But for myself, I cannot see the point in going to a doctor so that they can look for something that might be wrong when there is no indication right now of any problem. I have aches and pains that are only to be expected at my age and I don't need to go to a GP for an in-depth, six-minute consultation to be told I am overweight and should take more exercise. So I let my husband provide me with a single aspirin every day, and that is it.

My grandmother always warned me to stay away from hospitals. In her experience, going into hospital only increased your chances of coming out feet first in a pine overcoat. I don't want my life to be hampered by the constraints of a diagnosis or prognosis, to be defined by an illness or to become a medical statistic. Ultimately it is fate that will determine how long I live and when I die. I don't need my death to be prevented. We all have different opinions and temperaments and how far down the line we go to stave off illness and death must be a personal decision. Mine will probably be to wait until whatever it is that eventually takes me becomes critical. My preference is not to allow my dying and my death to be medicalised.

My life has been full. It has had some purpose. It has been

fun. I have met many wonderful people. My husband is my best friend. We have beautiful children and grandchildren. I have outlived my parents. Even if my original, more conservative, life expectancy still holds true, I still have seventeen years to go, and frankly, every single day between now and then I consider a bonus. Of course I would like all this to last as long as possible, but my main desire is for my death to conform to the natural order of the cycle of life—in other words, I want to die before my children and my grandchildren. Having seen the pain and suffering of parents who have lost a child, I would not wish that torment on anyone.

Now that I have less time in front of me than I have behind me, I am starting to focus on that threshold I must cross some time within the next thirty years. I am not afraid to cross it on my own. Indeed, I think I would rather die alone—privately, quietly, on my terms and at my pace. I don't want to be distracted by having to worry about the pain and grief of my loved ones. I would like to ensure that I have all my ducks in a row. I don't want to leave work or trouble for anyone else. I want it to be tidy and neat and the next logical step in my life. I don't want to be any bother.

So how would I like it to happen? If I do not want my dying to be like my father's, I would welcome the same kind of death: simply turning my face to the wall when I am ready. I don't believe I would have the bravery to kill myself and I must therefore be prepared to wait with some patience for death to arrive. Might I take the assisted dying pill if it were available? Perhaps, in certain circumstances, but I would not have the same kind of courage as Arthur, my trainee cadaver. I have great faith that society will come to its senses before I shuffle off this mortal coil and allow us to plan our death rather than endure it at the hands of well-meaning medical or care staff. I would like my exit to be natural: I don't want transplants, or CPR, or drip-feeding, or, in my final moments, a syringe full of opiates. Of course, I may be utterly deluding myself. It's quite possible that when the first little bit of pain creeps in, I will be shouting for the morphine. I doubt it, though. I don't like losing sensation or control. And I have always had a very high pain tolerance (three babies, no pain relief). Only time will tell if I am right. When death comes for me, I would like to be properly alive to have my personal conversation with her unencumbered by pharmaceuticals.

While Uncle Willie's death was pain-free, I think it was just a bit too swift for my liking. I don't want to die in my sleep, either. I view death as my final adventure and I am reluctant to be cheated out of a moment of it. I am only ever going to experience it once, after all. I want to be able to recognise death, to hear her coming, to see her, to touch her, smell her and taste her; to undergo the assault on all of my senses and, in my last moments, to understand her as completely as is humanly possible. This is the one event that my life has always been leading up to, and I don't want to miss anything by not having a front-row seat.

Perhaps I will be fortunate enough to die like Sir Thomas Urquhart, the well-travelled seventeenth-century polymath, writer and translator from Cromarty in the north-east of Scotland, who was declared a traitor by Parliament for his part in the royalist uprising at Inverness. He wasn't subjected to any particularly harsh penalties, although he was later held in the Tower of London and at Windsor for fighting on the royalist side in the Battle of Worcester. Urquhart was eccentric in the extreme. Among his claims was that his 109-times great-grandmother, Termuth, was the woman who found Moses in the bulrushes and his 87-times great-grandmother was the Queen of Sheba. After he was released by Oliver Cromwell, he returned to the Continent. It is said that on receiving news of the restoration of King Charles II to the throne, he laughed himself to death. Micromorts meet micromirths—what a way to go.

I doubt that will be my destiny, more's the pity. But I have a prediction for you. I think I will die before I am seventy-five. I suspect it will be heart-related, and as deaths from myocardial infarction peak on Mondays, at 11am, apparently, I am booking mine in for a Wednesday at noon.

Obviously I don't actually know how to die, having never done it before. But surely it can't be that difficult: everyone who has ever lived before me seems to have managed it well enough, with some possible exceptions among the winners of those tongue-in-cheek Darwin awards, who have all succeeded in bringing about their own deaths in ludicrous ways. I can't rehearse for it and I can't seek advice from anyone who has done it. So really there is no point in worrying about it. But I know I won't be alone. Whether or not there are others present, death will be with me, and she has more experience than anybody, so I am certain that she will show me what to do.

I imagine my death as being akin to yielding to permanent general anaesthetic. Everything goes black, you know no more and that is it, you are dead. If all there is beyond death is darkness, I won't be able to remember it anyway, which is a great shame. But perhaps this is all there is to it: a fugacious moment tacked on to the end of a long story like a final full stop.

I do, however, have some very definite plans for the being dead phase. I want to ensure that my body is put to full use for anatomical education and research and so I will bequeath my remains to a Scottish anatomy department. If I had a choice, I would rather be dissected by science students than medics or dentists, since I try to avoid doctors, and nobody likes going to the dentist, do they? For me, being the next Henrietta to a student anatomist would complete the circle of my life. I currently hold an organ-donor card and aim to sign my bequeathal forms on my sixty-fifth birthday, if I am spared. By then, the chances of the organs I have abused for so long being of any value to a living person will be pretty slim.

Tom is not happy. He doesn't want me to be dissected. Despite being an anatomist himself, he is charmingly old-fashioned and would like to see me treated to a quiet and respectful funeral and then laid to rest in a place where the girls can visit me, should they ever wish to. If I go first, it is likely he will get his way because I would never want to force him into doing anything that would cause him distress. However, if he goes first, I will scrupulously observe his wishes for his own death and then ensure that my own are clearly and tidily set out.

Ideally, I would like to be dissected in my own dissecting room, but I accept that it might be unfair on my staff to have to undertake the embalming process. They are professionals, and I imagine they would be absolutely fine with it, especially if it was my express wish, but I wouldn't want to risk upsetting any of them. However, I do want to be Thieled, and at present Dundee is the only place where this is possible. Becoming a formalin cadaver does not appeal and I absolutely refuse to be fresh/frozen. I like the idea of having that bit of extra flexibility to my limbs—probably more than they currently have—which Thiel provides and I would welcome the smoothing out of my wrinkles. And I'd be able to repose for a couple of months in the dark, cool waters of my submersion tank, enjoying a nice rest after all that dying nonsense. I wonder what aberrations in my anatomy will have some student somewhere cursing me one day, and whether I will be as good a teacher as Henry was to me.

Once everything has been dissected, I'd like my skeleton to be macerated (boiled to remove all the soft tissue and to get rid of the fat). I am happy for my soft tissue and organs to be cremated, though they won't leave much in the way of ash for my children to scatter. I have other plans for my bones. I want them to be stored in a box in the skeletal teaching collection at Dundee University. I will leave a full history of identifying features—injuries, pathology and so on—that can be related back to them. I would be just as delighted to be articulated and hung up in the dissecting room, or in our forensic anthropology teaching lab, so that I can continue to teach there long after I have stopped functioning. As bones have a very long shelf life, I could be hanging around for centuries, whether my students like it or not.

If I achieve my aim, I will never really die, because I will live on in the minds of those who learn anatomy and fall in love with its beauty and logic, just as I did. This is the kind of immortality we can all aspire to achieve in our own spheres. I would have no desire to live on in corporeal form for ever, even if I believed it were possible.

Some choose to reject the inevitability of complete death. Many are convinced that their soul, spirit or the essence of their identity will live on in some way, on earth or in their concept of heaven, despite the expiry of their body. Others believe their spirit will one day be reunited with its own body. Or, in the case of those who embrace the idea of reincarnation, in somebody else's. There are even a few people who have their bodies cryogenically frozen, until such time as medical science works out how to bring them back to life again just as they were before. None of this is for me.

Is there life after death? Who knows. And are there such things as ghosts? My superstitious grandmother would certainly have said so but, having spent much of my life around the dead, I can categorically state that no dead body has ever hurt me, and rarely has one offended me. The dead are not unruly, but generally very well behaved and polite. None of them has ever come back to life in my mortuary and they certainly do not haunt my dreams. All in all, the dead are a whole lot less trouble than the living. There is only one way to discover the truth about dying, death and being dead, and that is to do it, which we will all get round to eventually. I only hope I am ready and have my bag packed for the big adventure.

What does my heaven look like? Let's lose the angels and the harps—how irritating they would be. My heaven is peace, silence, memories and warmth.

And my hell? Lawyers, blue wires and rats.

THE MAN FROM BALMORE

Please contact Missing Persons at **missingpersonsbureau@nca.x .gsi.gov.uk** if you believe you have information that may help to identify the young man whose story is told in Chapter 8. A case history is available on **http://missingpersons.police.uk/en/case/11-007783**

DESCRIPTION OF REMAINS

Remains found: 16 October 2011. Likely to have been there for between 6 and 9 months

Location: Woodland near Golf Course Road, Balmore, East Dunbartonshire

Sex: Male

Age: Between 25 and 34

Ancestry: Northern European; fair hair

Height: Between 1.77m and 1.83m (5ft 8ins and 6ft)

Build: Slight frame

Distinctive characteristics: Injuries that may have affected his appearance: healed broken nose, which might have been visibly crooked; partly healed serious fracture of the jaw; chipped upper front tooth. Possible difficulty walking

CLOTHING

■ **POLO SHIRT**: Light blue, short sleeves, V-neck. White printed design of text and stamps covering the whole of the front. Dark-coloured integrated diagonal strip from the right top shoulder to the left hem

Brand: Topman. Widely available in the UK

Size: Small, Euro size 48; chest size 35-37

Fabric: 100% cotton

Manufacture and other labelling: Made in Mauritius. Label contains numbers with the sequences 2224278117026 and 71J27MBLE

■ CARDIGAN: Dark blue, long sleeves, crew neck, front zip, side pockets. Two horizontal stripes on collar and at top of pockets. 'SOUTHERN CREEK PENNSYLVANIA' embroidered on left breast above white crown, lion and the letters 'G' and 'J'. 'RIVIERA ADVENTURE' logo underneath

Brand: Max. Apparently traded only in the Middle East

Size: Small

Fabric: 100% cotton

Manufacture: Made in Bangladesh

■ JEANS: Denim, button fly

Brand: Petroleum. This brand offered both an 'essentials' range (Petrol-eum '68) and young fashion (Petroleum '79) at affordable prices in the UK. Sold in the UK exclusively at Officers Club, Petroleum stores and online

Size: 30L

Fabric: 78% cotton, 22% polyester

Other labelling: Petroleum, 'Don't blame me I only work here'

■ BRIEFS: Coloured boxers, red elastic waistband printed continuously with 'Urban Spirit'

Brand: Urban Spirit is a medium-priced brand commonly sold in the UK

■ TRAINERS: Laced, black and grey, red sole.'Shock X' on side. 'Rubber grip', 'Flex Area', 'Performance' and 'Brake' imprinted on sole

Brand: Some research suggests the logo

is registered to the German brand Crivit Sports, which was sold widely in Lidl and other budget stores

Size: 45/11

Fabric: 100% polyester

■ SOCKS: Nondescript, dark-coloured ankle socks

ACKNOWLEDGEMENTS

Reflecting on a lifetime of events always carries the risk of missing out someone terribly important and inadvertently causing offence. So I will simply thank every precious companion who has travelled with me on life's bus. Some were there for a stop or two; others have gone the entire distance alongside me. And what a road trip we have had. I don't need to reel off your names because you know who you are, and you know in your hearts how much you mean to me. I treasure your company, your friendship, your wisdom and your kindness.

If I have forgotten something, or have perhaps told a story not quite as you remember it, forgive me. And if our experiences together are not recalled on these pages, it may be because I feel they are too personal to share, or that there is insufficient space to do them justice. I take full responsibility for my failings.

While my life rolls on, the production of this book has been finite and I would like to thank those who have been endlessly patient with me, so encouraging, refreshingly honest and supportive.

Michael Alcock, above all others, has shown the patience of a saint. Having first listened to my ramblings over twenty years ago, he has finally seen something appear in print. I am very lucky to have found him and I adore him.

Caroline North McIlvanney knows better than anybody else that I have no words to thank her adequately for the Herculean task she accepted and then executed with such sensitivity and grace.

And Susanna Wadeson was inordinately brave to take on an amateur writer she heard speaking at a conference. She has been the most inspirational, comforting, reassuring and firm guide throughout this adventure. Without her, this project would never have come to fruition and my family are indebted to her for en-abling these stories to be told. She is truly remarkable.

My sincere thanks are due, too, to Patsy Irwin (publicity director), Geraldine Ellison (production manager), Phil Lord for the page design and Richard Shailer for the jacket design.

Finally, I wish to extend my respects to the unidentified man represented on the front cover of this book. He has no name because he is a construct of Richard's immense artistic talent. But even he can be brought to life—just a little. We know he is male from the acute angle of his sub-pubic concavity, the shape of the pelvic inlet, the relative size of the alae to the body width of the sacrum, the triangular shape of the pubic bone and the acute morphology of the greater sciatic notch. He is over twenty-five because the bodies of S1 and S2 have fused, as have his iliac crest epiphyses. He is likely to be under thirty-five, since there is no evidence of osteophytic lipping on the ventral margins of his lumbar vertebrae and no obvious calcification into the costal cartilages.

Talk about showing off.

PICTURE CREDITS

Facial reconstructions of Rosemarkie Man (p117) and the man from Balmore (p167) are reproduced courtesy of Dr Chris Rynn.

The photograph of Sue in Kosovo (p225) was taken by David Gross.

The cartoon (p307) is by Zebedee Helm.

The portrait of Sue (p327) is by Janice Aitken.

Photographs of the man from Balmore's clothing (pp340–1) were taken by Dr Jan Bikker.

All other illustrations are from Sue's own collection.

INDEX

Page numbers in *italics* refer to pages with illustrations.